HOW TO ACTUALLY TELEPORT THROUGH DIMENSIONS AND VISIT OTHER WORLDS

By Johnny Vincento

Johnny Vincento

HOW TO ACTUALLY TELEPORT THROUGH DIMENSIONS AND VISIT OTHER WORLDS

BY Johnny Vincento

UNITED STATES COPYRIGHT OFFICE

FIRST EDITION 2005

SECOND EDITION 2020

ISBN9798609953995

Library of Congress

COVER PHOTOS ARE FROM NASA

HOW TO ACTUALLY TELEPORT THROUGH DIMEN...

And Thomas said to Jesus:

"You have certainly persuaded us, Lord. We realize in our
Heart, and it is obvious, that this is so, and that your word is suffient. But these words that you speak to us are ridiculous and contemptible to the world since they are misunderstood. So how can we go preach them, since we are not esteemed in the world?"

The Savior answered and said, "Truly I tell you that he who will listen to your word and turn away his face or sneer at it or smirk at these things, truly I tell you that he will be handed over to the ruler above who rules over all the powers as their King, and he will turn that one around and cast him from Heaven down to the abyss,..."

The book of Thomas the contender:

Johnny Vincento

The Enclosed 20 years of Work, is the most complete research ever performed on the **"Multiverse Theory,"** the **"String Theory," and "Teleportation."** Teleportation can now be obtained. Religion and science are one! Many scientist absolutely deny any religious references. They are mostly atheists and for good reason at first glance, if you think about it. However, anyone who looks at the "Bible" or the "Emerald Tablets of Thoth" or some of the "Gnostic Gospels:" Will see many strange unexplained events. I am a "Pioneer Physicist" studying "Ancient Religious knowledge." These ancient writings seem absurd and ridiculous at first glance. When one looks at them from a Scientific light then one can see how these things can be. When one activates a **"Wormhole"** and sees and walks on another physical planet. Then one will understand that other worlds have different laws of physics. The **"Theory of Matter"** is the other half to complete the **"String Theory."** You will be able to visit other "Immortal Worlds" through this lost now found knowledge. For the world is at least 4 in one. And for that matter, Even 500 years before Jesus, Buddha was trying to cross dimensions to obtain pure truth which he called "Nirvana." Buddha didn't do the formula correctly, His method took way to long to open up the **"Silver Cord."**

What is the "Silver Cord?" The "Silver Cord" is also known as the life thread, connection or tube from the higher dimensions down to the physical body. This is also described as the connection or cord that joins your physical body to your Father in Heaven. Your "Silver Cord" is one that keeps your connection with your Creator down to your physical form and your Spirit can move back and forth through this connection to other worlds. In other words the "Silver Cord" can be used as a "Wormhole."

Included is the correct formula which has been tried over and over again with success every time.

Now "Teleportation" is available to Humanity. I will show you how to see with your own eyes and then you can believe what is written in these pages. This work describes in detail my personal experiences to three other physical worlds, These reside in overlapping universes. I didn't see the planets as a whole but visited them. So you understand, the "Multiverse Theory" already agrees on this theory. All the equations the physicists make say it's true. **Any equations one**

makes cannot disprove this as fact. The ancient writings that refer to traveling through dimensions and to other worlds are as follows: The "King James four Gospels" in certain verses, the "Gospel of Thomas," the "Gospel of Phillip", and the "Emerald Tablets of Thoth." My research says that the "Emerald Tablets" were written about 12,500 years ago or around 38,000 years ago. It is one or the other no inbetween. I lean toward the 10,500 BC date because, in the tablets Thoth talks about he came to Egypt and built the marker in his image (The Sphinx) after Atlantis sank. **"Plato"** describes Atlantis sinking 9,000 years before him. So that would put more evidence for the 12,500 year dating for the Emerald Tablets. Not to mention, the new dating on the Sphinx due to water erosion and the alignment of the constellation "Leo."

Through harnessing ones own spirit one opens a "Wormhole." In the "Old Testament" this is called "The Silver Cord." Thoth describes it as a **"Silver Spark."** This is what was used to visit other overlapping worlds during Jesus's time and Thoth's time. Those two are the all time enlightened ones. That word "Silver Cord" is a religious word. However, what is trying to be described is a "Wormhole." The "Wormhole" is not a hole, because **until now,** in our time, no one has gone through one except those with near death experiences. The Wormhole is a tube. That's why many near death experiences describe the **"Silver Cord"** as going through a **Tube** with light at the end of the **Tunnel.** The powers of creation itself can be harnessed by an individual. **For the spirit is the only thing that can go through dimensions.**

Without a doubt the impossible will not be believed. Few will believe, because they cannot see. Now you will know how to see and still you won't believe your own eyes. Only when you understand what you are seeing will you believe.

Those who live for the physical will not have any part of the next discovery of life. Do not bother with those. Only those who have spirits that are crying in pain, or ones who want to push science, will attempt to see. For those people your spirits will rejoice. Jesus said **"Don't give pearls to swine, for they will trample on them and rend you."** Rend means attack.

Some of the issues that are found as fact in this work are as follows: The Earth is a physical mortal one and within dimensions combined with at least three other overlapping physical immortal ones. Unseen due to dimensions and accessed through "Teleportation." Thoth wrote that **"Nine are the worlds within worlds."**

Some of the subjects that will be analyzed in detail are as follows: The formula to visit other physical planets (immortal overlapping duplicated Earths). Activating the formula when outside Earths gravitational pull:(A possible living overlapping Mars and unknown planets with simular life in the galaxy or universe) may be found. The fact that ones who are Paralyzed will walk in the other physical realms. **"The Theory of Matter," which explains how "Teleportation" is possible.** The "Theory of Time Travel," "Seeing the Future (Get prepared for 2030AD) Seeing the Past, and many more.

The enclosed formula is meant to ensure your "Teleportaiton." The fatigue needed can be obtained in many ways. I only used liquor one hour ahead. One can probably obtain it with just lack of sleep alone. For that matter, exercise should bring the desired extreme fatigue needed also. So you do not have to partake in liquor if you wish not to.

• •

HOW TO ACTUALLY TELEPORT THOUGH DIMENSIONS AND VISIT OTHER WORLDS

I
Teleportation
Journey to my Father
A Message from another Dimension
The Gospel of Philip from the Dead Sea Scrolls
The Emerald Tablets of Thoth:
written around 12,500 yrs ago
describe Atlantis, Teleportation, Sphinx,
and the Pyramids of Giza

II
The Theory of the Miracle
If my Dog Can Walk You Can Walk
The Father, the Son and You
Children of Cancer
The Parable of the Blind Man: from John Vincento

III
The Gospel of Thomas from the Dead Sea Scrolls
The Theory of Time Travel
The Theory of Seeing the Future
Visiting the Dead

IV
Accessing Creation
The Birth of the Spiritnaut
A Creature Extinct is Alive
A Talk with Immortals (Level Two)

V

The Purity of the Heart
Paradise Turns to Terror
Paradise Can Be Paradise

VI

Natural Laws of Interdimensional Teleportation
Children of Heaven
✦✦The Theory of Matter

VII

✦✦Investigation into the Resurrection of Jesus
Father How Long Will I Live?

VIII

Sayings of the Spiritnaut John Vincento
Nostradamus: Only a Mystic Can Understand a Mystic
The Alpha and the Omega

IX

Letter to the Unborn of the 22nd Century:
The Realm Boundary and the "Dimension Separation Viewer"
Letter to the Unborn of the 22nd Century:
The Theory of Bending Space by the Rematerialization
of Matter Through Dimensions
Letter to Nasa
Letter to the Reader
Letter to Catholic Parents
Letter to the People of Asia: Buddhists, Hindus, Christians

X

Final Thoughts from the Author
Not Everyone Makes It (Song)
letter to all the suicide killers and evil ones
The Third Heaven

XI

HOW TO ACTUALLY TELEPORT THROUGH DIMEN...

Vision from the Spirit of Truth "2030 A.D."

I

TELEPORTATION: YOU CAN OBTAIN THIS WHILE YOU LIVE

The world doesn't understand, you are not resurrected in this world when you die. People don't know, you are resurrected in another physical world. When you acquire it while you live, your one body will sleep on Earth, and your spirit will enter into your physical second body of flesh: At this point through the "Theory of Matter," ones own body is duplicated on another physical planet that is immortal. So for the duration of the Journey you are two physical bodies.

Both worlds occupy the same space, separated by dimensions, and accessed through the:

1. "Resurrection" when you die which is a (one way trip)

Johnny Vincento

or

2. "Teleportation" which is a safe proven (two way trip)   while you live.

Jesus said, "On the day when you were one, you became two. But when you became two, what will you do?" Gospel of Thomas 11

This is a guide to perform the lost art of "Teleportation." I made it possible for individuals to see with their own eyes. Do not take anyone's word for the truth. I don't want you to believe my words, I want you to believe your own eyes: Then believe my words. Furthermore, Gospel passages are implemented into the experiences I have seen. The individual is to make their own conclusions.

"Those who say they will die first and then rise again are in error. If they do not first receive the resurrection while they live, when they die they will receive nothing." Gospel of Philip Paragraph 73

(Here is described taking a **"Teleportation Journey"** while a person is still alive instead of waiting to die. The reason not described in this verse but elsewhere. Is that, he describes **the spirit needing to secure the path** or in scientific terms: Secure the "Silver Cord" or "Wormhole" **between physical worlds.** Apparently the older a person gets the weaker their "Silver Cord" gets. Thus, for this reason a persons spirit can't go through their (broken tube) "Wormhole" and they get stuck as a spirit on this physical Earth. Instead of having their body duplicated in physical form across dimensions on one of the Immortal planets.

This being accomplished through the **"Theory of Matter" combined with** the **"String Theory."** Making one complete theory.

Jesus said, "If you do not fast from the world you will not find the Kingdom." Gospel of Thomas 27

Jesus said, "The Father's Kingdom is spread out upon the earth and people don't see it." Gospel of Thomas 113

"His disciples said to him, "When will the rest for the dead take place, and when will the new world come?" He said to them, "What you are looking forward to has come, but you don't know it." Gospel of Thomas 51

<u>**Attain the "Teleportation" in the Summer on a Sunday**</u>

<u>CONSULT A DOCTOR BEFORE YOU FAST & FATIGUE! EVERYONE IS DIFFERENT AND MAY BE DANGEROUS IF YOU'RE NOT HEALTHY! DO SO AT YOUR OWN RISK!</u>

Johnny Vincento

FORMULA FOR THE TELEPORTATION
**This is the THREE DAY PLAN
for those who are thin or petite:**
9am Thursday – 9am Friday: Fast and sleep.
9am Friday – 9am Saturday: Fast and sleep.
9am Saturday – 9am Sunday: Fast
and do not sleep.
At 8am on Sunday sip on 80 proof liquor
to obtain great
fatigue only.
At 9am close your eyes and sleep: immediately
you will be in your resurrected body of flesh
in one of the overlapping physical worlds

FORMULA FOR THE TELEPORTATION
**This is the FOUR DAY PLAN
for those who are of average weight:**
Same as above except add:
9am Wednesday – 9am Thursday: Fast and sleep.

The body must be made weak enough to visit the other worlds. You only have up to 5 minutes there, so it is a recommended idea to have a friend wake you after exactly 4 minutes and 45 seconds. This is so you do not feel as you will be stranded there. **Also, do not have your friend touch you around your mid section: There is an invisible "Silver Cord" which is what the "Wormhole" is. It is attached to both physical worlds and both physical bodies. It is attached to round about the**

center of your chest.

This formula must be adapted to your individual body. It is very close, but not exact. If you activated the formula and did not receive the Journey, then the next day repeat the fasting and fatigue: That should accomplish it. Read "Journey to my Father," I lost my nerve and no longer wanted to take the Journey: But it was too late, the formula was already activated.

Being a " Pioneer Physicist," I tell you exactly what was done to attain the "Teleportation." However, the "Teleportation" cannot be guaranteed. Those who are not healthy: Being overweight for example. Fasting will never weaken the body enough, because their body will still obtain nutrition.

WHEN YOU COME BACK FROM YOUR JOURNEY START TO EAT SLOW. FOR EXAMPLE A LITTLE EVERY HOUR. BY THE END OF THE DAY YOUR SYSTEM WILL BE NORMAL.

Johnny Vincento

JOURNEY TO ANOTHER WORLD THROUGH "THE SILVER CORD WORMHOLE"> JOURNEY TO MY FATHER (LEVEL 1)

You were born with a Father on Earth, but you have another Father in Heaven: Who you can talk to in person.

Fasting for three days, I sipped on 80 proof liquor for courage. For I know, the time has arrived for my trip to the unknown world. And courage is indeed needed, for the Journey is taken only by the boldest of individuals. With liquor providing courage, it also provides the needed fatigue that is combined with the fasting on the third day. This accesses the secret of the world: <u>For the world is two in one!</u>

On the third night, I drank past the point of about to fall asleep. I lost my nerve at that time and instead of falling asleep and undergoing the Journey: I continued on drinking to avoid it. The next morning I woke up weary and that afternoon, with no liquor, I fell asleep at some time past noon and this was a Sunday.

Immediately I was standing in my resurrected body, in the home that occupied the same space as mine: In the other world. Before me was a sight not to be taken easily: A single file line of standing dead women. <u>I say dead because they were dead and I</u>

<u>say they were alive with spirit too!</u> They were gray, drained of blood, unable to function, except for a small walking shuffle. The first woman was in her forties, with shoulder length blond hair parted in the middle. Her eyes were black because they were dilated open all the way. She was wearing a white see through nightgown and weighed about 120lbs at about 5ft 7in tall. The second was black haired with the right side of her face totally mangled. So much so, it could not even be recognized as a face. The other side of her face was intact.

The rest I avoided eye contact. Passing them, I looked back at them from behind and started to bang my head against the wall: To feel pain so to get back to my other body. I did not feel any pain, only pressure. I entered the last room to my left and seen my own Father: Looking through chest high stacked cardboard boxes. It was good to see a friendly face. I went to the window of the empty room and looked out. The second story view was strange to me, since I knew this home occupied the same space. How come the street cul-de-sac was the same shape, but covered in crushed white rocks? How come there is a horse corral across the street? There is only fields without a home in sight? In my world all these fields are full of homes! <u>THEN I SPOKE TO THE ONE WHO WAS IN THE LIVING RESURRECTED IMAGE OF MY OWN FATHER ON MY EARTH.</u>

This is what was said:

Johnny Vincento

"Father, who are those people in the hallway?"

"Those are dead people."

"Well some of them look like they have been in car wrecks, where am I?"

"You are here."

"No I'm not, I'm at home sleeping."

"How could you be at home sleeping if you are here?"

"Listen, I'm telling you, I'm not here, I'm at home sleeping in bed."

"Cannot you understand the difference, between being at home sleeping in bed and standing here, in front of me, in this room?"

"I don't care what you say, I need to wake up, I want you to slap me in the face!"

"Are you sure?"

"Yes."

(Then I was slapped front hand and back hand)
"Harder again!"

(Then I was slapped the same, with no pain, only impact pressure)
"I need to get out of this house!"

"You will never make it past the dead people."

"Do you have any weapons?"

"No."

"Not even a kitchen knife to open your boxes?"

"No, I do not have any weapons."

"Well I'll take my chances!"

Once my Father said I would never make it past the dead people, it was clear to me that this was not my Father and a trick of that world. For he did not sound the same or use the same vocabulary. For I thought at that time my Father is alive in my world: How can he be in Heaven? <u>These are the things I thought at that time, for I did not understand yet, that I have two Fathers: One on Earth and one in Heaven.</u>

I then thought to myself I'll rush the hallway while their backs are turned, then turn around after passing them, so they can't jump on my back. I charged the hallway expecting a fight, I turned to face them and I was walking backwards. Their walking shuffle was getting faster. I visualized the kitchen knife I wish I had because the living dead were closing in. Then the knife I thought of appeared out of thin air, on the hall room white rug. I picked it up and it was tangible and made of stainless steel. I could feel the steel. I was about to stab the blond dead woman that was in the lead, but I had a hard time telling the difference between this world and that one. The worlds are exactly the same physical wise: <u>You can't even tell the difference!</u> I did not want to get charged with murder, but then thought to myself, "This isn't my world, anything goes here!" The blond woman had the ability to float, the last five feet or so she did not walk, she *seemed to glide toward me. Then I was face to face with the dead woman. I'm telling you that 4 inches from my face I looked into the living deads eyes. I panicked, dropped the knife, and ran down the stairs. I could hear my feet banging down the stairs. The floor plan was the same, so I knew how to get out of the house. I grabbed the brass push button handle and rushed out of the house.*

I then tripped on high ground during my mad rush out and fell in the grass. The grass was the greenest grass that I have ever seen: I did not see one dead blade, the grass was about five inches high. I could feel the grass and then I rolled over on my back. I stretched my arms out and became intoxicated with pleasure, with the sight of the sky and clouds on that beautiful day. It was as if my emotions were multiplied times ten. I thought to myself, "This is the first time I have ever been outside in this world. I want to see how far this world goes beyond the house itself."

I thought if I turn around and see the horse corral across the street, that will prove that the sight from the upstairs window was not a false one. I turned around and seen a joyfull sight: The brown horse, stone road, wooden fence were all there, just like I seen from the window.

<u>**I said out loud**</u>, "This isn't just a house, it's a whole other world!"

I got up and stood in the grass facing the house and

it looked different from my world.

I yelled "Father get down here now, you got some explaining to do!"

He said: "I will be down in five minutes." ***I yelled out*** "Five minutes is too long, I want you down here in 30 seconds!"

He replied, "Ok" ***and I awoke in my bed, lying on my side. The weather too was beautiful in both worlds that day.***

UPDATE: This was written in the earliest of research.

A MESSAGE FROM ANOTHER DIMENSION

In the summer of the year 2003, there came a message from Heaven.

In the early afternoon, in broad daylight, I touched up my finished woodcarving. The summer before, I underwent "Journey to my Father." Creating this carving was my tribute to all that I have seen.

While doing this I had these thoughts:

**"Father in Heaven, I hope you will
receive this carving.
For many things are duplicated**

> **in the second world.
> I know you live in the home that
> overlaps this one.
> I have seen many things and I know you wish
> me to write those things which I have seen.
> But it is too much! I don't know
> if I have the words!"**

Wide awake, sitting with my tribute, a
vision came to me from the North East.
On the wall, in the air, toward the ceiling,
words started to form.
The letters formed slowly, as if put together
with grains of black sand.
When complete they said:

Thou hast

The letters were about three inches high and rounded at ends. I turned away, because I didn't believe what I had just seen. So I expected to look again and see the words gone. However, I looked again and saw an amazing sight: The words were still there to behold! I thought to myself: "Father I will try to do your will." Then after about ten seconds more, the words started to dissolve: As if disappearing one grain of sand at a time. Then my spirit rejoiced, because someone in Heaven loves me.

Note: For about 6 months after this vision I

thought that "Thou hast" meant: **"I have."** **Meaning that he received my carving.**

Someone told me then:
"Thou hast" meant: **"You have."** **Meaning that I have the words!**

If it wasn't for this vision, I would have never put together such a complicated 20 years of research.

In the Old testament **Daniel 5:5 "In the same hour came forth fingers of a man's hand, and wrote over against the candlestick upon the plaister of the wall of the king's palace: and the king saw the part of the hand that wrote."** I did not see any hand or fingers when the words were formed in my wide awake vision.

THE GOSPEL OF PHILIP
From the Gnostic Gospels
Summary of One Hundred Paragraphs

Those who say that the Lord died first and then rose up are

in error, for he rose up first and then died. If one first attains the resurrection, he will not die.
Paragraph 16

It is through water and fire that the whole place is purified – the visible by the visible, the hidden by the hidden. There are some things hidden through those visible. There is water in water; there is fire in chrism.
Paragraph 18

And so he dwells either in this world or in the resurrection or in the middle place. God forbid that I be found in there! In this world, there is good and evil. Its good things are not good, and its evil things not evil. But there is evil after this world, which is truly evil – what is called "the middle." It is death. While we are in this world, it is fitting for us to acquire the resurrection, so that when we strip off the flesh, we may be found in rest and not walk in the middle. For many go astray on the way. For it is good to come forth from the world before one has sinned.
Paragraph 49

Those who say they will die first and then rise are in error. If they do not first receive the resurrection while they live, when they die they will receive nothing.
Paragraph 73

The world came about through a mistake. For he who created it wanted to create it imperishable and immortal. He fell short of attaining his desire. For the world never was imperishable nor, for that matter, was he who made the world. For things are not imperishable, but sons are. Nothing will be able to receive imperishability if it does not first become a son. But he who has not the ability to receive, how much more will he be unable to give?
Paragraph 79

Knowledge of the truth merely makes such people arrogant, which is what the words, "it makes them free" mean. It even gives them a sense of superiority over the whole world.

But "love builds up" (1 co 8:1), in fact, he who is really free, through knowledge, is a slave, because of love for those who have not yet been able to attain to the freedom of knowledge. Knowledge makes them capable of becoming free. Love never calls something its own.
Paragraph 87

If anyone becomes a son of the bridal chamber, he will receive the light. If anyone does not receive it while he is here, he will not be able to receive it in the other place. He who will receive that light will not be seen, nor can he be detained. And none shall be able to torment a person like this, even while he dwells in the world. And again when he leaves the world, he has already received the truth in the images. The world has become the aeon (eternal realm), for the aeon is fullness for him. This is the way it is: it is revealed to him alone, not hidden in the darkness and the night, but hidden in a perfect day and a holy light.
Paragraph 100

THE EMERALD TABLETS OF THOTH

TABLET 9: SUMMARIZATION
"Know you that throughout the space that you dwell in are others as great as your own. Interlaced through the heart of Matter yet separated in space of their own."

"Once in a time long forgotten. I, Thoth, opened the doorway. Penetrated into other spaces and learned of the secrets concealed."

"Deep in the essence of Matter are many mysteries concealed. Nine are the interlocking dimensions and nine are the cycles of space."

"Nine are the levels of Consciousness and nine are the worlds within worlds. Space is filled with concealed ones. For space is divided by time. Seek you the key to the Time Space and you shall unlock the gate."

Know you that throughout the Time Space. Consciousness's surely exist. Though from our knowledge it is hidden. Yet it still forever exists."

"The key to worlds within thee are found only within. For man is the gateway of mystery and the key that is one with the one."

"You are the key to all wisdom. Within you is all Time and Space."

TABLET 10: SUMMARIZATION
"Listen you man, and hear a mystery stranger than all that lies beneath the sun."

"Know you man, that all space is filled by worlds within worlds, yes, one within the other yet separated by law."

TABLET 11: SUMMARIZATION
"Listen you now while I give you wisdom. Use it and free you shall be.

"Know you that in the pyramid I built are the keys that shall show you the way into life."

"Yes, draw you a line from the great image I built, to the apex of the pyramid, built as a gateway."

"Draw you another opposite in the same angle and direction. Dig you and find that which I have hidden."

"There shall you find the underground enterance to the secrets hidden before you were men."

"Yes in a time that is yet unborn,...Man, a perfect flame of this cosmos, shall move forward to a place in the stars. <u>Yes, shall move even from out of this Space Time into another beyond the stars.</u>"

TABLET 1: SUMMARIZATION: THOTH DESCRIBES WHAT HAPPENED TO

ATLANTIS

"I, Thoth, the Atlantean, master of mysteries, keeper of records. mighty king, magician..."

"Great were my people in the ancient days. Great beyond the conception of the little people now around me. Knowing the wisdom of old, seeking far within the heart of infinity. Knowledge that belonged to earths youth."

"Then having drunk deep of the cup of wisdom, I looked into the hearts of men and there found I greater mysteries and was happy. For only in search for truth could my soul be stilled and the flame within me be quenched."

"Over the world then broke the great waters, drowning and sinking, changing the entires earths balance. Until only the temple of light was left standing on the great mountain on Undal which was still rising out of the water. some there were who were living, saved from the rush of the fountains."

TABLET 5: SUMMARIZATION: THOTH THINKS BACK AND DESCRIBES THE EVENTS FURTHER

"Then gathered I the sons of Atlantis, into the spaceship I brought all my records. Brought the records of sunken Atlantis gathered I all of my powers. Instruments many of mighty magic up then we rose on Wings of the Morning." (Name of their flying craft?)

"High we arose above the temple,..."

"Fast fled we then on the Wings of the Morning, fled to the land of the Children of Khem." (Khem is the earliest Egypt)

"There by my power I conqued and ruled them. Raised I to light, the Children of Khem. Deep beneath the rocks I buried my space ship. waiting for the time when man might be free."

"Over the space ship I erected a marker. In the form of a lion yet a man. There beneath the image rests my space ship. To be brought forth when need shall arise."

"Deep beneath the image lies my secret. Search and find in the pyramid I built. Each to the other is the keystone, each the gateway that leads into life."

"Follow the key I leave behind me. Seek and the doorway to life shall be yours. Seek you in my pyramid deep in the passage that ends in a wall.

Johnny Vincento

TABLET 1: SUMMARIZATION

"Raised over the passage, I a mighty pyramid. Using the power that overcomes Earth Force ..."(Anti gravity?)

"A circular passage reaching almost to the great summit."

"There in the apex, set I the crystal, sending the ray into the Time Space. Drawing the Force from out of the ether, concentrating on the gateway to Amenti." (Amenti is a different planet that Thoth visits through Teleportation. Immortal people live there)

"Other chambers I built and left vacant to all seeming. Yet, hidden within them are the keys to Amenti." (Most likely the air vents and the empty sarcophagus)

"He who in courage would dare the realms, <u>let him be purified first by long fasting.</u>"

NEXT

<u>"Lie in the sarcophagus of stone in my chamber. Then I will reveal to him the great mysteries. Soon shall he follow to where I shall meet him."</u>

If one activates their "Silver Cord Wormhole," in the Great Pyramid of Giza, in the inner chamber within the sarcophagus. **At least one air shaft must be open. For they are not only air shafts but an aiming device to travel to other galaxies through the art of "Teleportation."** The "Silver Cord" is sent through the air vent aiming at the desired star system. Thoth describes everyone he met on other worlds as immortal. No matter what planet it was. For that reason, and from my Journeys as well. The spirit bends space when crossing dimensions. This is described in great detail within "The Theory of Bending Space by the Rematerialization of Matter through Dimensions."

Some of the keys mentioned, should be small square stone blocks to either open one air vent and close another or vice versa. Depending to what star system you want to go. When the one pyramid with the sarcophagus was built the air vents were aimed at different star systems. One aimed toward Amenti)

"Built I the great pyramid, patterned after the pyramid of Earth Force, burning eternally so that it too might remain through the ages."

"In it I built my knowledge of magic science..."

TABLET 4: SUMMARIZATION
"Abstaining from eating until you have conquered desire for food, that is bondage of soul."

NEXT

"Then lie down in the darkness. Close your eyes from the rays of the light. Centre your soul force in the place of your consciousness, shaking it free from the bonds of the night."

NEXT

"Place in your mind place the image you desire to go. Picture the place you desire to see. Vibrate back and forth with power."

NEXT

"Loosen the soul from out of its night. fiercely must you shake with all your power. Until at last your soul shall be free."

(Back then humans had more senses than ours. My formula bypasses the vibration by adding extreme fatigue instead)

Recap: Around 1300 BC in Alexandria the Emerald Tablets were translated. The creation date of 12,500 for these writings are for the following reasons. The "Precession of the Equinoxes" is the rotation of the earth as it makes one complete turn. Found by 2nd century BC astronomer "Hipparchus."

This axis rotation makes a complete rotation in approximately 26,000 years. Or present day calculations are more exact at 25,772 years.

The story of Atlantis was written in about 360 BC by the Greek philosopher "Plato." it can be found in his "Timarus" and "Critias" dialogues.

According to the dialogues: "Critias" told how his grandfather had met an Athenian poet who told of the Egyptian story about Atlantis.

This is from "Timacus:"
"Let me begin by observing first of all; that nine thousand was the sum of years which had elapsed..."

"Many great deluges have taken place during the nine thousand years; for that is the number of years which have elapsed since the time of which I am speaking of."

Next is the constellation of "Leo." which would be in front of the Sphinx 12,500 years ago. The lion man would be looking at it. Others date the Sphinx at 25,772 years plus 12,500 years, which is why they say it could be from around 38,000 years as well. It can only be from one time period or the other.

Thus, the earth would have to make one complete turn to put the constellation "Leo" back where it was again. in front of the Sphinx when it was created.

The water erosion on the Sphinx that would take thousand of years continues the proof of it's age. This research is one more piece of proof for the 12,500 age. Even though "Plato" said 9,000 years that was an estimate. Close enough.

<u>The one pyramid was built as a gateway to teleport across dimesions and visit other worlds.</u>

II

THE THEORY OF THE MIRACLE

"And as Jesus passed by, he saw a man
which was blind from birth
and his disciples asked him, saying,
Master, who did sin, this man,

> or his parents, that he was born blind?
> Jesus answered, neither hath
> this man sinned, nor his parents: But
> that the works of God should
> be made manifest in him. I must work
> the works of him that sent me,
> while it is day: The night cometh, when
> no man can work." JN 9: 1-4

"The miracle is a natural law of physics of the other realm, falling under the "Law of function." This theory comprises the following four elements:

1. Healing oneself while in the second world
2. Healing others while in the second world
3. Attempting to heal oneself while in this world
4. Attempting to heal others while in this world

HEALING ONESELF WHILE IN THE SECOND WORLD

It is not opinion that one paralyzed will be able to walk while visiting the other realm. Under the "Law of function," nerve damage and related internal injuries are not transferred to ones second body. The following Journey tells us many things:

JOURNEY TO ANOTHER WORLD THROUGH "THE SILVER CORD WORMHOLE"> IF MY DOG CAN WALK YOU CAN WALK (LEVEL 1?)

My spirit crushed, my spirit crying in pain, the third night befell me. I drown myself in 80 proof one hour before it was time. Fasting for days, I cursed out the world for destroying my Great Dane Lanie. Born to just an average life span of 8 years. What kind of world is this? First giving my poor dog cancer so she can't walk, taking away the use of both rear legs, making her a cripple. Luckily I was body building, so I was able to carry her everywhere she wanted. Then she died and my spirit felt like it died too! Now it is time to light the candles.

I can feel the fatigue of the glass as I place the flickers to be my beacon home. When you come back to a dark room you won't panic, because the candles will let you know you are home. For both worlds are identical as far as the senses go, it is difficult to tell the difference. My eyes heavy, my mind on my dog, I prayed nothing because my spirit was crushed. I set down my 80 proof glass with tears in my eyes and say out loud "Lanie it's me!"

Immediately, I was standing in the front door hallway, in the house that occupies the same space: In the other world. Before me in front of the same staircase as my house, was my Great Dane. This is what I said:

"Snoopy, it's me, come here."
"I thought you would be able to walk on all four legs in the spirit world."
"Well three out of four isn't bad."
"I miss you Snoopy."
"I'm sorry for teasing you."
"I was just playing; didn't I always let you know after teasing you that I was just joking?"
"Didn't I always pet you to let you know that I was not mad at you and just playing?"

I kneeled down and hugged her and I could feel that unique Great Dane fur: And she was of flesh. I then returned to my body and my spirit rejoiced!!! When I see her again I am going to let her know that her rear right leg is healed too. Both worlds feel exactly equal, so she doesn't know her leg is fine. This is why believing is so important.

Anyone who attains the "Teleportation" while they live will enter their second body. They will be of the same image including physical damage to the flesh. This being the "Law of duplication" and the "Law of transfer." However, through the "Law of function," you will be healed. Anyone with Parkinson's should be healed, those born blind should see, and those who are paralyzed will walk.

HEALING OTHERS WHILE IN THE SECOND WORLD

Just as my dog thinks she can't walk, so there should be many in the immortal realm that do not believe, or do not understand. These ones are healed already, but since they do not believe, they are still affected by their ailments that they had in the mortal world. When I see my dog again, I am going to handle her leg until she understands that both legs are fine. It goes to say, if one leg of hers is healed, then the other one must be healed also.

As in the four Gospels of King James, one can heal others exactly like Jesus healed. Jesus brought the natural laws of the overlapping world to this world. That realm has different laws of physics, matter and the mind, which manipulates function and matter. It is most important to inform them of what this lesson is. As long as they believe they will be healed. However, I do not pick and choose verses. That is why I add this difficult verse to this element:

> "Then came the disciples to Jesus apart and said why could not we cast him out?
> And Jesus said unto them, because of your unbelief:
> For verily I say unto you,
> If ye have faith as a grain of mustard seed, ye shall say unto this mountain, remove
> Hence to yonder place; and it shall remove: And nothing shall be impossible unto
> you. Howbeit this kind goeth not out but by

prayer and fasting." MT 17: 19-21

This verse can be related to both worlds. Casting a demon out of a person in an overlapping world, will take belief on both parties. The same for the mortal realm we reside in. Jesus is referring to this world though. Moving a mountain may be possible in the second world, due to the "Theory of matter," which is another law of physics in that realm. Since he had harnessed these new laws of nature from the other world, the impossible now is possible in our world. Moving a mountain in the Bible most likely means "a problem in a persons life." From what I experienced in these other physical realms your mind is very powerful: And probably could move an actual mountain.

ATTEMPTING TO HEAL ONESELF AND OTHERS IN THIS WORLD
FIRST ATTEMPT:
Attain the "Teleportation" through the formula and make sure you have a friend to wake you after five minutes. Make sure this is a nice Sunday morning in the Summer. When you find yourself duplicated across dimensions in the flesh, call out for your Father in Heaven. When you find him <u>*TELL HIM*</u> "Father in Heaven, I want you to heal my other body, so when I go back to my world I will be able to walk." When you come back, stand up right away. You must believe, if your Father in Heaven tells you that you are now

healed, that would really make you believe. We do not know what he will say though. Make sure you have other questions memorized to ask him. What ever you see over there remember, nothing can harm you.

SECOND ATTEMPT:
Be it so, that one may be healed in Heaven and attempt to bring the healing gift of the Father to Earth. Along with this, the powers of healing others and ones own ailment, may be possible in our realm of mortal existence.

Jesus it seems continuously fasted for days on end. All the way up to forty days at a time. Only eating enough to sustain himself. This coming to a point when the resurrection is attained everyday! A balance of sustainable nutrition, fasting and fatigue. Thereby combining both worlds and in doing so, bringing the natural laws of Heaven to Earth.

There is a point when Heaven and Earth are open to one another: And the individual is the door. This experience is described in "Accessing Creation." At that moment for instance, if a paralyzed person was in the room with me, it may have been possible to heal him. This coming to be only if he believed it was possible. If that one in the wheel chair did this himself, the natural laws of the other world would be open to him and he may be able to heal himself in our world.

This experience may also "charge" an individual

for a limited period of time, after the Journey: Giving healing powers.

MANY OF THE MIRACLES JESUS PERFORMED WERE ON THE SABBATH DURING THE DAY *THIS BEING THE SABBATH BEFORE CHRISTIANITY CHANGED IT TO SUNDAY, THE JEWISH SATURDAY SABBATH*

"The healing at the pool" JN 5: 9-10
"A crippled woman healed" LK 13: 10-17
"Jesus heals a man with an unclean spirit" MK 1: 21-28
"Jesus heals Simon's mother in law" MK 1: 29-31
"Jesus heals the man who had dropsy" LK 14: 1-6
"Jesus heals a man with a withered hand" MK 3: 1-6

Update: My dogs real name was Lanie, but in Heaven I called her by her nickname "Snoopy."

THE FATHER, THE SON, AND YOU

The following may be difficult to comprehend. This due, to the subject always being focused on one person: Jesus, he being the Son of his Father in Heaven. I tell you, his real teachings of this complex issue have been misunderstood and lost to time.

THE FATHER

You the individual, each have a Father in Heaven. He takes the image of your own Father on Earth. You have the ability to see and talk to him in person. This being accomplished, through the "Teleportation" while you are still alive.

> "In that hour Jesus rejoiced in spirit, and said, I thank thee, O Father, Lord of Heaven and Earth, that thou hast hid these things from the wise and prudent, and hast revealed them unto babes: even so, Father; for so it seemed good in thy sight."

> "All things are delivered to me of my Father: and <u>no man knoweth who the Son is, but the Father; and who the Father is, but the Son, and he to whom the Son will reveal him.</u>"

> "And he turned him unto his disciples, and said privately, blessed are the eyes which see the things that ye see: for I tell you, that many prophets and kings have desired to see those things which ye see, and have not seen them; and to hear those things which ye hear, and have not heard them." Luke 10: 21-24

> Jesus said, "If you do not fast from the world, you will not find the Kingdom. If you do not observe the Sabbath as a Sabbath you will not see the Father." Gospel of Thomas: 27

> Jesus said, "When you see the one who was not born of woman, fall on your faces and worship. That one is your Father." Gospel of Thomas: 15

> Jesus said, "Congratulations to those who have been persecuted in their heart: they are the ones who

have truly come to know the Father. Congratulations to those who go hungry, so the stomach of the one in want may be filled." Gospel of Thomas: 69

The following paragraph is as heavy duty as it comes, concerning the Father.

Jesus said, "Images are visible to people, but the light within them is hidden in the image of the Fathers light. He will be disclosed, but his image is hidden by his light." Gospel of Thomas: 83

THE SON

If you attain the "Teleportation," while alive or after death: <u>You are the Son.</u> You become reborn in your new immortal body. You are then a child of the resurrection. Thus, being compared to a second birth.

Jesus said, "Whoever drinks from my mouth will become like me; I myself shall become that person, and the hidden things will be revealed to him." Gospel of Thomas: 108

CHILDREN OF CANCER

For you I write the following words; even though you are not yet even born. My Father in Heaven knows you will need to be comforted. I am the "Unique One," who has not forgotten about you. The ability to see the invisible is within me now. However, when you are born into this world, I will not be here to comfort you. So I send my words to you through time.

Be not afraid, for I will explain things. You have a Father in Heaven who loves you very much. He is your true Father. He created you, by you I mean your spirit, your existence. Your father on Earth helped create your body. So you have two fathers and the one in Heaven will be waiting for you! When you were in Heaven as a spirit you did not have a body. You were a sphere which looked like it was filled with liquid water. You did not have a face on your spirit, because you were not given a body yet. The only way your spirit could receive a body is to be born on Earth. Now see, you had to be born so your spirit could have a body. And now that you have a body, your spirit will be able to duplicate it in Heaven. That is the cycle of things. When you die you become alive. Your body will be duplicated exactly the way it is. Same age, same condition. Heaven is a physical place and your body that your spirit duplicates will be physical. So take care of it and if need be, live on Earth without medical treatments that destroy the body. Your Father in Heaven is real child. I promise you on my soul.

Your Heavenly Father looks exactly like your Dad on Earth. When you were born, you received a body and at the same time your Father in Heaven, who was a spirit also, received a body: Your Dads.

So I tell you the way it is. Take care that your body is not damaged, because it will be duplicated in the same condition. Do not become an Organ Donor, Do not become embalmed.

I know your spirit will be crying. Even though you are not born yet. This is the reason I write to you across time. You are sad about your existence ending and inside you cry out "I don't want to die" – "It's me." You must be born three times to complete your existence.

> 1. You are born in Heaven as a spirit.
> 2. You are born on Earth with a body.
> 3. You are born back in Heaven with a body for your spirit.

Then you become immortal completing the cycle of life.

Now for an example: When a newborn baby dies on Earth, that baby will age to eight years old in Heaven. Age in that realm is not viewed the same as here. So that baby's age will be an emotional one. Its body will grow until the inner age is that of an eight year old. There is the "Outer age" and the "Inner age." In Heaven, the child may appear nine or ten years old, but the physical aging stops when the "Inner age" is eight for that situation.

Comfort in my words for I give you this final example: Growing to be very elderly is not a good thing in the greater knowledge of things. Can you imagine walking around in Heaven in a hundred year old wrinkled body forever?

The same is for you, but the opposite: Can you imagine walking around in Heaven as a child forever? Which sounds like the better?

Johnny Vincento

THE PARABLE OF THE BLIND MAN
From John Vincento

There was a certain fellow who was born without sight. Now at the age of 19 he heard of such things that Jesus did. He asked many if this was true and they all said they did not know. Then he asked his parents and they said the same: But it was written.

Now this fellow was troubled by all this. For he would like to see as did the one's healed by Jesus. So he went to a church and asked the priest if there was a way to receive this miracle. The priest replied, "There is not a way, only Jesus can do such things. I am sorry I can not help you."

Alone in a park, the man cried with his head in his hands. A man passing said, "What troubles you so?" He replied "I have no sight since birth and the only one who can heal me has been gone for two thousand years. I'm afraid I will never see."

Little did he know the one he chanced across was a spirit healer. The stranger said, "There is a way."

After hearing this, he wiped his tears and said "Tell me so I may see."

He explained the formula to him and the spirit healer spoke these words: "The Journey you must take is taken alone and in secret. You will be the first to seek sight. For if the crippled can walk, so should too the blind at birth see. I give to you the lost mystic art of Jesus Christ. When you close your eyes you will be in a different place.

HOW TO ACTUALLY TELEPORT THROUGH DIMEN...

As soon as you arrive you should see for the first time. This will be a Sunday morning, so call out for your Father in Heaven. He will look like your own father and will be of flesh. Tell him to restore your sight when you return to your other body in this world."

"If anything, you will know your sight will await you and your name is written in Heaven."

III

THE GOSPEL OF THOMAS
From the Gnostic Gospels

These are the secret sayings that the living Jesus spoke and Didymos - Thomas recorded.

1. And he said, "Whoever discovers the interpretation of these sayings will not taste death."

2. Jesus said, "Those who seek should not stop seeking until they find. When they find, they will be disturbed. When they are disturbed, they will marvel, and will reign over all. And after they have reigned they will rest."

3. Jesus said, "If your leaders say to you, "Look, the Father's Kingdom is in the sky," then the birds of the sky will precede you. If they say to you, "It is in the sea, then the fish will precede you. Rather, the Kingdom is within you and it is outside you."

"When you know yourselves, then you will be known, and you will understand that you are children of the living Father. But if you do not know yourselves, then you live in poverty, and you are the poverty."

4. Jesus said, "The person old in days won't hesitate to ask a little child seven days old about the place of life, and that

person will live. For many of the first will be last, and will become a single one."

5. Jesus said, "Know what is in front of your face, and what is hidden from you will be disclosed to you. For there is nothing hidden that will not be revealed. And there is nothing buried that will not be raised."

6. His disciples asked him and said to him, "Do you want us to fast? How should we pray? Should we give to charity? What diet should we observe?"
Jesus said, "Don't lie, and don't do what you hate, because all things are disclosed before Heaven. After all, there is nothing hidden that will not be revealed, and there is nothing covered up that will remain undisclosed."

7. Jesus said, "Lucky is the lion that the human will eat, so that the lion becomes human. And foul is the human that the lion will eat, and the lion still will become human."

8. And he said, "The person is like a wise fisherman who cast his net into the sea and drew it up from the sea full of little fish. Among them the wise fisherman discovered a fine large fish. He threw all the little fish back into the sea, and easily chose the large fish. Anyone here with two good ears had better listen!"

9. Jesus said, "Look, the sower went out, took a handful of seeds, and scattered them. Some fell on the road, and the birds came and gathered them. Others fell on rock, and they didn't take root in the soil and didn't produce heads of grain. Others fell on thorns, and they choked the seeds and worms ate them. And others fell on good soil, and it produced a good crop: it yielded sixty per measure and one hundred twenty per measure.

10. Jesus said, "I have cast fire upon the world, and look, I'm guarding it until it blazes."

11. Jesus said, "This Heaven will pass away, and the one above it will pass away. The dead are not alive, and the living will not die. During the days when you ate what is dead,

you made it come alive. When you are in the light, what will you do? on the day when you were one, you became two. But when you become two, what will you do?"

12. The disciples said to Jesus, "We know that you are going to leave us. Who will be our leader!" Jesus said to them, "no matter where you are you are to go to james the just, for whose sake Heaven and earth came into being."

13. Jesus said to his disciples "Compare me to something and tell me what I am like."

Simon peter said to him, "You are like a just messenger." Matthew said to him, "You are like a wise philosopher."

Thomas said to him, "Teacher, my mouth is utterly unable to say what you are like."

Jesus said, "I am not your teacher. Because you have drunk, you have become intoxicated from the bubbling spring that I have tended."

And he took him, and withdrew, and spoke three sayings to him. When Thomas came back to his friends they asked him, "What did Jesus say to you?"

Thomas said to them, "If I tell you one of the sayings he spoke to me, you will pick up rocks and stone me." (the original scroll missing here or fragmented or is too extreme for children, or a back ground check of original words retranslated along with first two lines of 14).

14. "When you go into any region and walk about in the countryside, when people take you in, eat what they serve you and heal the sick among them. After all, what goes into your mouth will not defile you; rather, it's what comes out of your mouth that will defile you."

15. Jesus said, "When you see one who was not born of woman, fall on your faces and worship. That one is your Father."

16. Jesus said, "Perhaps people think that I have come to easy peace upon the world. They do not know that I have come to cast conflicts upon the earth: fire, sword, war."
"For there will be five in a house: there'll be three against two and two against three, father against son and son against father, and they will stand alone."

17. Jesus said, "I will give you what no eye has seen, what no ear has heard, what no hand has touched, what has not arisen in the human heart."

18. The disciples said to Jesus, "Tell us, how will our end come?"
Jesus said, "Have you found the beginning, then, that you are looking for the end? You see, the end will be where the beginning is."
"Congratulations to the one who stands at the beginning: that one will know the end and will not taste death."

19. Jesus said, "Congratulations to the one who came into being before coming into being."

"If you become my disciples and pay attention to my sayings, these stones will serve you. For there are five trees in paradise for you; they do not change, summer or winter, and their leaves do not fall. Whoever knows them will not taste death."

20. The disciples said to Jesus, "Tell us what Heaven's kingdom is like." He said to them, "It's like a mustard seed, the smallest of all seeds, but when it falls on prepared soil, it produces a large plant and becomes a shelter for birds of the sky."

21. Mary said to Jesus, "what are your disciples like?" he said, "they are like little children living in a field that is not theirs. When the owners of the field come, they will say, "give us back our field." They take off their clothes in front of them in order to give it back to them, and they return their field to them. For this reason i say, if the owners of a house know that a thief is coming, they will be on guard before the thief arrives and will not let the thief break into their house (their domain) and steal their possessions."

"As for you, then, be on guard against the world, prepare yourselves with great strength, so the robbers can't find a way to get to you, for the trouble you expect will come. Let there be among you a person who understands."
"When the crop ripened, he came quickly carrying a sickle and harvested it. Anyone here with two good ears had better listen!"

22. Jesus saw some babies nursing. He said to his disciples, "These nursing babies are like those who enter the Kingdom." They said to him, "Then shall we enter the Kingdom as babies?"

Jesus said to them, "When you make the two into one, and when you make the inner like the outer and the outer like the inner, and the upper like the lower,"..."When you make eyes in place of an eye, a hand in place of a hand, a foot in place of a foot, an image in place of an image, then you will enter the Kingdom."

Jesus said, "I shall choose you, one from a thousand and two from ten thousand, and they will stand as a single one."

24. His disciples said, "Show us the place where you are, for we must seek it." He said to them, "Anyone here with two ears had better listen! There is light within a person of light, and it shines on the whole world. If it does not shine, it is dark."

25. Jesus said, "Love your friends like your own soul, protect them like the pupil of your eye."

26. Jesus said, "You see the sliver in your friend's eye, but you don't see the timber in your own eye. When you take the timber out of your own eye, then you will see well enough to remove the sliver from your friend's eye."

27. "If you do not fast from the world, you will not find the Kingdom. If you do not observe the Sabbath as a Sabbath you will not see the Father."

28. Jesus said, "I took my stand in the midst of the world, and in flesh I appeared to them. I found them all drunk, and I did not find any of them thirsty. My soul ached for the children of humanity, because they are blind in their hearts and do not see, for they came into the world empty, and they also seek to depart from the world empty. But meanwhile they are drunk. When they shake off their wine, then they will change their ways."

29. Jesus said, "If the flesh came into being because of spirit, that is a marvel, but if spirit came into being because of the body, that is a marvel of marvels."

"Yet I marvel at how this great wealth has come to dwell in this poverty."

30. Jesus said, "Where there are three deities, they are divine. Where there are two or one, I am with that one."

31. Jesus said, "No prophet is welcome on his home turf; doctors don't cure those who know them."

32. Jesus said, "A city built on a high hill and fortified cannot fall, nor can it be hidden."

33. Jesus said, "What you will hear in your ear, in the other ear proclaim from your rooftops. After all, no one lights a lamp and puts it under a basket, nor does one put it in a hidden place. Rather, one puts it on a lamp stand so that all who come and go will see its light."

34. Jesus said, "If a blind person leads a blind person, both of them will fall into a hole."

35. Jesus said, "One can't enter a strong person's house and take it by force without tying his hands. Then one can loot

his house."

36. Jesus said, "Do not fret, from morning to evening and from evening to morning, about your food, what you're going to eat, or about your clothing, what you are going to wear. You're much better than the lilies, which neither card nor spin."

"As for you, when you have no garment, what will you put on? Who might add to your stature? That very one will give you your garment."

37. His disciples said, "When will you appear to us, and when will we see you?" Jesus said, "when you strip without being ashamed, and you take your clothes and put them under your feet like little children and trample them, then you will see the son of the living one and you will not be afraid."

38. Jesus said, "Often you have desired to hear these sayings that I am speaking to you, and you have no one else from whom to hear them. There will be days when you will seek me and you will not find me."

39. Jesus said, "The Pharisees and the scholars have taken the keys of knowledge and have hidden them. They have not entered nor have they allowed those who want to enter to do so. As for you, be as sly as snakes and as simple as doves."

40. Jesus said, "A grapevine has been planted apart from the Father. Since it is not strong, it will be pulled up by its root and will perish."

41. Jesus said, "Whoever has something in hand will be given more, and whoever has nothing will be deprived of even the little they have."

42. Jesus said, "Be passersby."

43. His disciples said to him, "Who are you to say these

things to us?" "You don't understand who I am from what I say to you. Rather, you have become like the Judeans, for they love the tree but hate its fruit, or they love the fruit but hate the tree."

44. Jesus said, "Whoever blasphemes against the Father will be forgiven, and whoever blasphemes against the son will be forgiven, but whoever blasphemes against the Holy Spirit will not be forgiven, either on earth or in Heaven."

45. Jesus said, "Grapes are not harvested from thorn trees, nor are figs gathered from thistles, for they yield no fruit."

"Good persons produce good from what they've stored up; bad persons produce evil from the wickedness they've stored up in their hearts, and say evil things. For from the overflow of the heart they produce evil."

46. Jesus said, "No one is so much greater than John the Baptist."

"But I have said that whoever among you becomes a child will recognize the Kingdom and will become greater than John."

47. Jesus said, "A person cannot mount two horses or bend two bows. And a slave cannot serve two masters, otherwise that slave will honor the one and offend the other."

"Nobody drinks aged wine and immediately wants to drink young wine. Young wine is not poured into old wineskins, or they might break, and aged wine is not poured into a new wineskin, or it might spoil. An old patch is not sewn onto a new garment, since it would create a tear."

48. Jesus said, "If two make peace with each other in a single house, they will say to the mountain, "move from here!" and it will move."

49. Jesus said, "Congratulations to those who are alone and chosen, for you will find the Kingdom. For you have come from it, and you will return there again."

50. Jesus said, "If they say to you, "Where have you come from?" say to them, "We have come from the light, from the place where the light came into being by itself, established itself, and appeared in their image."

If they say to you, "Is it you?" say, "We are its children, and we are the chosen of the living Father."

If they ask you, "What is the evidence of your Father in you?" Say to them, "It is motion and rest."

51. His disciples said to him, "When will the rest for the dead take place, and when will the new world come?" He said to them, "What you are looking forward to has come, but you don't know it."

52. His disciples said to him, "Twenty four prophets have spoken in Israel, and they all spoke of you." He said to them, "You have disregarded the living one who is in your presence, and have spoken of the dead."

53. His disciples said to him, "Is circumcision useful or not:" He said to them, "If it were useful, their father would produce children already circumcised from their mother. Rather, the true circumcision in spirit has become profitable in every respect."

54. Jesus said, "Congratulations to the poor, for to you belongs Heaven's Kingdom."

55. (original scroll damaged here, fragmented or to extreme for children)

56. Jesus said, "Whoever has come to know the world has discovered a carcass, and whoever has discovered a carcass, of that person the world is not worthy."

57. Jesus said, "The Father's Kingdom is like a person who has good seed. His enemy came during the night and sowed weeds among the good seed. The person did not let the workers pull up the weeds, but said to them, "No, otherwise you might go to pull up the weeds and pull up the wheat

along with them." "For on the day of the harvest the weeds will be conspicuous, and will be pulled up and burned."

58. Jesus said, "Congratulations to the person who has toiled and has found life."

59. Jesus said, "Look to the living one as long as you live, otherwise you might die and then try to see the living one, and you will be unable to see."

60. He saw a samaritan carrying a lamb and going to Judea. He said to his disciples, "That person {destroyed here} around the lamb." They said to him, "So that he may kill it and eat it." He said to them, "He will not eat it while it is alive, but only after he has killed it and it has become a carcass." They said, "Otherwise he can't do it." He said to them, "So also with you, seek for yourselves a place for rest, or you might become a carcass and be eaten."

61. Jesus said, "Two will recline on a couch; one will die, one will live." Salome said, "Who are you mister? You have climbed onto my couch and eaten from my table as if you are from someone." Jesus said to her, "I am the one who comes from what is whole. I was granted from the things of my Father." "I am your disciple."
"For this reason I say, if one is whole, one will be filled with light, but if one is divided, one will be filled with darkness."

62. Jesus said, "I disclose my mysteries to those who are worth of my mysteries."

"Do not let your left hand know what your right hand is doing."

63. Jesus said, "There was a rich person who had a great deal of money." He said, "I shall invest my money so that I may sow, reap, plant, and fill my storehouses with produce, that I may lack nothing." These were the things he was thinking in his heart, but that very night he died. Anyone here with two ears had better listen."

64. Jesus said, "A person was receiving guests. When he had prepared the dinner, he sent his slave to invite the guests. The slave went to the first and said to that one, "my master invites you." That one said, "some merchants owe me money; they are coming to me tonight. I have to go and give them instructions. Please excuse me from dinner." The slave went to another and said to that one, "My master has invited you." That one said to the slave, "I have bought a house, and I have been called away for a day. I shall have no time." The slave went to another and said to that one, "My master invites you." That one said to the slave, "My friend is to be married, and I am to arrange the banquet. I shall not be able to come. Please excuse me from dinner." The slave went to another and said to that one, "my master invited you." That one said to the slave, "I have bought an estate, and I am going to collect the rent. I shall not be able to come. Please excuse me." The master said to his slave, "Go out on the streets and bring back whomever you find to have dinner." Buyers and merchants will not enter the places of my Father."

65. He said, "A person owned a vineyard and rented it to some farmers, so they could work it and he could collect its crop from them. He set his slave so the farmers would give him the vineyard's crop; they grabbed him, beat him, and almost killed him, and the slave returned and told his master. His master said, "Perhaps he didn't know them." He sent another slave, and the farmers beat that one as well. Then the master sent his son and said, "Perhaps they'll show my son some respect." Because the farmers knew that he was the heir to the vineyard, they grabbed him and killed him. Anyone here with two ears had better listen!"

66. Jesus said, "Show me the stone that the builders rejected: That is the keystone."

67. Jesus said, "Those who know all, but are lacking in themselves, are utterly lacking."

68. Jesus said, "Congratulations to you when you are hated and persecuted; and no place will be found, wherever you have been persecuted."

69. Jesus said, "Congratulations to those who have been persecuted in their heart: They are the ones who have truly come to know the Father."
"Congratulations to those who go hungry, so the stomach of the one in want may be filled."

70. Jesus said, "If you bring forth what is within you, what you have will save you. If you do not have that within you, what you do not have within you will kill you."

71. Jesus said, "I will destroy this house, and no one will be able to build it."

72. A person said to him, "Tell my brothers to divide my father's possessions with me." He said to the person, "Mister, who made me a divider?" he turned to his disciples and said to them, "I'm not a divider, am I?"

73. Jesus said, "The crop is huge but the workers are few, so tell the harvest boss to dispatch workers to the fields."

74. He said, "Lord, there are many around the drinking trough, but there is nothing in the well."

75. Jesus said, "There are many standing at the door, but those who are alone will enter the bridal suite."

76. Jesus said, "The Father's Kingdom is like a merchant who had a supply of merchandise and found a pearl. That merchant was prudent; he sold the merchandise and bought the single pearl for himself. So also with you, seek his treasure that is unfailing, that is enduring, where no moth comes to eat and no worm destroys."

77. Jesus said, "I am the light that is over all things. I am all: from me all came forth, and to me all attained. Split a piece

of wood; I am there. Lift up the stone, and you will find me there."

78. Jesus said, "Why have you come out to the countryside? To see a reed shaken by the wind? And to see a person dressed in soft clothes, like your rulers and your powerful ones? They are dressed in soft clothes, and they cannot understand truth."

79. A woman in the crowd said to him, "Lucky are the womb that bore you and the breasts that fed you." He said to her. "Lucky are those who have heard the word of the Father and have truly kept it. For there will be days when you will say, "Lucky are the womb that has not conceived and the breasts that have not given milk."

80. Jesus said, "Whoever has come to know the world has discovered the body, and whoever has discovered the body, of that one the world is not worthy."

81. Jesus said, "Let one who has become wealthy reign, and let one who has power renounce."

82. Jesus said, "Whoever is near me is near the fire, and whoever is far from me is far from the Kingdom."

83. Jesus said, "Images are visible to people, but the light within them is hidden in the image of the Father's light. He will be disclosed, but his image is hidden by his light."

84. Jesus said, "When you see your likeness, you are happy. But when you see your images that came into being before you and that neither die nor become visible, how much you will have to bear!"

85. Jesus said, "One who came from great power and great wealth, is not worthy of you. For had he been worthy, he would not have tasted death."

86. Jesus said, "Foxes have their dens and birds have their nests, but human beings have no place to lay down and rest."

87. Jesus said, "How miserable is the body that depends on a body, and how miserable is the soul that depends on these two."

88. Jesus said, "The messengers and the prophets will come to you and give you what belongs to you. You, in turn, give them what you have, and say to yourselves, "When will they come and take what belongs to them?"

89. Jesus said, "Why do you wash the outside of the cup? Don't you understand that the one who made the inside is also the one who made the outside?"

90. Jesus said, "Come to me, for my yoke is comfortable and my lordship is gentle, and you will find rest for yourselves."

91. They said to him, "Tell us who you are so that we may believe in you." He said to them, "You examine the face of Heaven and earth, but you have not come to know the one who is in your presence, and you do not know how to examine the present moment."

92. Jesus said, "Seek and you will find. In the past, however, I did not tell you the things about which you asked me then. Now I am willing to tell them, but you are not seeking them."

93. "Don't give what is holy to dogs, for they might throw them upon the manure pile. Don't throw pearls to pigs, or they might (destroyed here) it (destroyed here)."

94. Jesus said, "One who seeks will find, and for one who knocks it will be opened."

95. Jesus said, "If you have money, don't lend it at interest. Rather, give it to someone from whom you won't get it back."

96. Jesus said, "The Father's Kingdom is like a woman. She took a little leaven, hid it in dough, and made it into large loaves of bread. Anyone here with two ears had better lis-

ten!"

97. Jesus said, "The Father's Kingdom is like a woman who was carrying a jar full of meal. While she was walking along a distant road, the handle of the jar broke and the meal spilled behind her along the road. She didn't know it; she hadn't noticed a problem. When she reached her house, she put the jar down and discovered that it was empty."

98. Jesus said, "The Father's Kingdom is like a person who wanted to kill someone powerful. While still at home he drew his sword and thrust it into the wall to find out whether his hand would go in."

99. The disciples said to him, "Your brothers and your mother are standing outside." He said to them, "Those here who do what my Father wants are my brothers and my mother. They are the ones who will enter my Father's Kingdom."

100. They showed Jesus a gold coin and said to him, "The Roman emperor's people demand taxes from us." He said to them, "Give the emperor what belongs to the emperor, give god what belongs to god, and give me what is mine."

101. "Whoever does not love father and mother as I do cannot be my disciple. For my true mother gave me life."

102. Jesus said, "damn the Pharisees! They are like a dog sleeping in the cattle manger: The dog neither sleeps nor lets the cattle eat."

103. Jesus said, "Congratulations to those who know where the rebels are going to attack. They can get going, collect their imperial resources, and be prepared before the rebels arrive."

104. They said to Jesus, "Come, let us pray today, and let us fast." Jesus said, "What sin have I committed, or how have I been undone? Rather when the groom leaves the bridal suite, then let people fast and pray."

105. {original scroll damaged, fragmented, or is too extreme for children}

106. Jesus said, "When you make the two into one, you will become children of the resurrection, and when you say, "Mountain, move from here!" it will move."

107. Jesus said, "The Kingdom is like a shepherd who had a hundred sheep. One of them, the largest, went astray. He left the ninety nine and looked for the one until he found it. After he had toiled, he said to the sheep, "I love you more than the ninety nine""

108. Jesus said, "Whoever drinks from my mouth will become like me; I myself shall become that person, and the hidden things will be revealed to him."

109. Jesus said, "The Father's Kingdom is like a person who had a treasure hidden in his field but did not know it. And when he died he left it to his son. The son did not know about it either. He took over the field and sold it. The buyer went plowing, discovered the treasure, and began to lend money at interest to whomever he wished."

110. Jesus said, "Let one who has found the world, and has become wealthy, renounce the world."

THE THEORY OF TIME TRAVEL

Where are the dead of the ages who obtained the resurrection? There are many who claim to have seen ghosts dressed from a different time. This happens when the veil of dimensions lifts up for a moment.

Johnny Vincento

You must remember, the second world occupies the exact same space, but separated by dimensions. They say the ghosts can be seen through. They only appear that way from our world. <u>They are the resurrected and are of flesh.</u> The difference is that they are immortal. You cannot walk through a wall in the second world. In "Journey to my Father," I was banging my head against the wall. If a ghost walks through a wall, it is because in their world the floor plan is without a wall. Say there is a home from the 1880's and a ghost is seen. She is dressed in the clothes of the time. When she died in the 1880's, her clothes were resurrected along with her body and her aging stopped. In your time, you wish to make contact. Your city may be modern, but hers might have stayed as it was in the 1880's. All the homes you see in our world may be cornfields in hers.

Obtain the "Teleportation" in that home and make sure you pray to see what you are seeking. You should be able to meet her and shake her hand. Have questions in mind to ask her. Look out the windows and see the differences. Remember to have no fear, because nothing can harm you. You only have five minutes so make your questions good. Ask for her name.

My personal experience is written in "Journey to my Father." I was in the home that occupied the same space as mine. The neighborhood was hardly developed yet. There were empty fields as far as I could see toward the West. In this world it is full of homes. The cul-de-sac street is blacktopped here, but there, the cul-de-sac and street were the same shape but covered in white crushed rocks. Across the street instead of a home, there was a wooden fence with a brown horse behind it. The horse had a black main.

Many experiments can be done in the cities of antiquities. Pompeii, Ghost Towns of the West, the Aztec ruins. Obtain the "Teleportation" in these ruins: Those cities might still be booming! My theory is you may see amazing things traveling through time!

Update:

It came to be a fitting sight, for ten years past I have been traveling across dimensions to visit the resurrected. And now for the first time, one has come to visit me!

On the seventh of July of the year 2008, I was sitting in my chair. It was night out and my doors were shut. I had the lights on and in front of me about 12ft, appeared a man who materialized out of thin air. He was as solid as we look. I have a very weak "Duncan Fife" round end table. However, he was sitting on it with his right leg horizontal and bent down at the knee. His left leg was on the floor. He said "Are you looking for me?" and then disappeared out of thin air. His pants were black slacks, his shirt was a white dress type with buttons and cuffs at his wrists, however about 50% longer: Approx. 5 to 6 inches. The shirt was unfamiliar, because it was out of time. He was wearing shoes not boots and I only glanced at them: They were black.

This building may have been built in the 1930's, at least the library was and the construction is similar. The shirt col-

Johnny Vincento

lar was longer too. His hair was jet black, dry, length was about 3 to 4 inches round about: Reminded me of a pirate or a 1980's hair style with an unfamiliar white shirt. Months after, I seen on television a program about a surgeon from the 1930's. The photo shown, was a man wearing the same type of white dress shirt.

The visitors complexion was Caucasian. To further add, my only neighbor next to my apartment said he was living with a ghost. One example he told me: Was that at about 3am in the morning he was playing music loud on his computer. Then he said his chair was lifted about 4 or 5 inches off the ground and dropped. Now since my table could not have supported an approx. 150 pound person, there may be a table over my table in an apartment that overlaps mine.

THE THEORY OF SEEING THE FUTURE

Jesus said: "Congratulations to those who know where the rebels are going to attack. They can get going, collect their imperial resources, and be prepared before the rebels arrive. Gospel of Thomas 103

In the other overlapping worlds time should be able to be opened. To open time, one must seek out the Spirit of Truth. This is the most difficult theory to get started with: For one must first find the Spirit of Truth. Only that spirit can open time.

IT IS WRITTEN

"Even the Spirit of Truth; whom the world cannot receive, because it seeth him not; neither knoweth him: But ye know him; for he dwelleth with you, and shall be in you."
JN 14:17

The last line tells us, that through the resurrection, we can meet this spirit in person. This is the same theme as all my research. "The Kingdom is within you and it is outside you." Gospel of Thomas 3.

"Howbeit when he, the Spirit of Truth is come, he will guide you into all truth: For he shall not speak of himself; but whatsoever he shall hear, that shall he speak: And he will show you things to come." JN 16:13

To begin your search, seek out your Father in Heaven. Tell him what you want. Asking does not seem to work. You must use the words "I want." Tell him "Face to face" I want to see the future." If he cannot provide, then tell him "I want to talk to the Spirit of Truth." That is all you can do, to get this theory started. Update: Read Genisis 32:30

Johnny Vincento

"Face to face."

Update:

On the 26th day of March, of the year 2006, I did attempt to see the future. I told my Father in Heaven, "I want to see the future." He said "Not everyone has the ability to see the future." So I said, "I want you to show me the future." He then tested me to see if I possessed the gift. What was interesting is that he also tested me to see if I could see the past. I was unable to see the future, I did not concentrate, however I did see the past for about a quarter of a second. He did know me well, from my emotional age, which was correct, along with knowing I was a fan of the American Indians. Refer to "Father how long will I live?" to truly know how to see the future.

It is just one step away, to seeing if the average individual has the ability to see the future. The method for seeing the past may be able to be made without the presence of your Father in Heaven. When you find yourself in your resurrected body in Heaven, point your finger to a area of air that is outlined by objects. For example, any space of sky that is outlined by objects to make a complete frame. Then concentrate on what past experience you would like to see. Then you will be in that body in that time, for the vision is complete, as for you are there! As for the future, the only last step will be seeking the Spirit of Truth in person. Obtain the "Teleportation" on a Sunday and call out for the Spirit of Truth. If you see your Father in Heaven first, tell him that you want to speak with the Spirit of Truth. Tell your Father that you want to ask some questions about the future.

Even though I did not possess the ability to see the future, it is written that I need not this ability. That Spirit will tell me of what I ask, "He will show me things to come." JN 16:13, so I need not the gift.

The Journey is a long one to prepare for and a short one while you are there. So when you do find the Spirit of Truth through the "Teleportation" and you are shown things to come: Ask what date and year these things will come to pass.

If you are not given a date or year, then there is another way, whether it is accurate is unknown. It seems to be a favorite among the mystics of antiquity. The position of the Heavens must be documented. The day of your "Teleportation" can be dated in the Heavens. That date will come to pass again when the planets are lined up the same. However, unfortunately, the planets will be lined up the same numerous times in the future. The visions of future events that you are shown will occur on one of those alignment dates. That is why it is important to ask the date and year. Prepare yourself with specific questions you are going to ask. A good question for mankind would be: "What year will an asteroid hit the Earth?"

Update:

I have before me the planet alignments for March 26, 2006. This is the Journey that I was able to see my past: From "Father how long will I live?"
I received a toy when I was about 5 or 6 years old. The vision was me holding the toy. Some planets it seems are lined up around the first month of 1976. At that time I was 7 years old. That would be Mars, Venus and Saturn. All planets are never aligned. Prophets of antiquity never said all planets will be in certain constellations and that's when the vision will occur.

Update:
In the year 2013 The Spirit of Truth gave me a vision, that is documented at the end of this book: Concerning the year 2030 AD.

Johnny Vincento

VISITING THE DEAD

Now for the first time in history it is possible to visit the dead. This is more than just a theory, for I have seen and held my own dog.

The spirit can receive the resurrection while one lives through "Teleportation." For that matter, the spirit goes to a place one considers home. The spirit chooses that place to obtain the resurrection.

My dog died and now lives in the house that overlaps mine. Out of the ten "Teleportations" I have achieved as of now, I knew where I was five of those times. For even though I have been in the second world, the location of space is the same. Space is separated by only a hairs thickness between dimensions.

So I found myself at familiar places: The family business, the block I grew up at, and three times at the home I live in. The other five I did not know where I was: There was not time to explore.

In this age there was 9/11, but there was a girl who held the heart of the nation. Her name was Rachel. She died at the Columbine shootings in Littleton, CO. The way she was murdered and the things she said before she was shot, saddened me more than 9/11. Any family who lost a child due to murder or accident, I wrote this chapter for you.

If your loved one did receive the resurrection, there is a good guess where to find that individual. That person may be living in the home that overlaps the home they lived in. That is to say if they considered it a home. Obtain the "Teleportation" in that place.

Praying is thought by many, as being a waste of time. However, praying is very important when you are accessing creation itself. The spirit must know where you want to go and it takes along with it the mind. It also keeps your memories, that's why it is important to pray.

If you are lucky, you will find whom you seek in the flesh: Then your spirit will rejoice! If you don't find who you are looking for, do not be discouraged. The second world is an entire planet with many places to visit.

If one lost a loved one to murder, or accident, it would be fitting to acquire the "Teleportation" and heal your spirit.

When you seek out a child who died and find that person, how much will your spirit rejoice?

****
*

If your family lost a child, the child is most likely living in the home that overlaps yours. So making contact could be no further away than three or four days. Not only will your spirit rejoice if you find your child, you can ask specific questions to find who did the crime.

If one wishes to visit characters of history. Pray for whom you seek for two weeks first and then acquire the "Teleportation" in a place that the character would consider home.

IV

JOURNEY TO ANOTHER WORLD THROUGH "THE SILVER CORD WORMHOLE">ACCESSING CREATION

After the third day of fasting, night began to fall. I prepared to engage the final step of my "Silver Cord." As I sipped on my liquor, I readied myself to harness the powers of the Universe; and to walk within the parallel dual Earth as was written. The affects of the liquor come, but only severe fatigue is what I seek. Sleep arrives moments after I set down my 80 proof glass.

As soon as I close my eyes, I find myself physically sitting in a white 1980's pickup truck. I said to myself, "It worked." My body is in bed asleep but duplicated to make two! I pat my arms and legs and I am in another physical body. I feel the vinyl blue seats. As I sit in the passenger side, I look out the windshield and see the family business and warehouse. The warehouse is without any rust. I do not understand, a 360 degree view of madness. Another physical world occupying the same space, unseen due to dimension and accessed through "The Silver Cord." I can even smell the oil rags on the floor and I see the sky is dim.

Suddenly tear shaped spirits swarm the truck; flying all around. Their numbers were three or four. I reached to roll up the passenger side window and I could feel the round plastic handle. Then I stretched over to the drivers side, to roll up that window. The size of the spirits, were of a volleyball, stretched to a teardrop and the bottom of the tear was a human face.

Then I opened my eyes to my body in bed and sat up. Before my eyes, was my room, for about three seconds, then the view was from the pickup of the warehouse. This went back and forth about four or five times. It was as if I was jumping back and forth from my body on this Earth in my room, to my resur-

rected body in the pickup truck on the second Earth. The family business is about twenty miles away, so it seems space was bent, to right on top of my room. That world dimension, stretched for my ability to travel in Heaven while still alive on this Earth.

Then I was changing views to my room again and two spirits jumped to my world. One zoomed around my bed; the other then hovered over my head and shown its face. The face was flesh color and the tear was of the color of liquid water. Only the face was shown, no ears, hair, or scull. I punched at it wildly with both fists yelling, "Get away from me!"

Then the madness disappeared, as an image would, if it were made of sand and dropped all at once.

JOURNEY TO ANOTHER WORLD THROUGH "THE SILVER CORD WORMHOLE"> THE BIRTH OF THE SPIRIT-NAUT (LEVEL 1?)

Alone with my spirit, equal was the formula as the last. Heavenly and ghostly visions of which cannot be believed. It's time to see if the formula is perfected. Three days of fasting and my body is weak enough to send my spirit. I sip slowly on 80 proof liquor, I pray to see all things unseen. The liquor does its purpose. Sleep takes hold heavy, my eyes heavy, I say last out loud, "Spirit take me to the other world."

Sitting half upright with my arms crossed over my chest, I close my eyes to find myself immediately in my resurrected body. I was standing in a room, with a window before me, and a desk to my left. I say to myself "I made it, this is great, I am really here!" My joy was times ten. I patted my sleeveless arms and could feel my flesh. I then patted the front of my legs. I was wearing the same checkered cloth pajamas. They were duplicated with my body. I said to myself "I don't want any doubt what so ever," that is when I snapped the elastic on my pajamas to my skin. I then said out loud, "Yeap, I'm here." I felt the white see through drapes and it was made of cloth. I rubbed it in my fingers. I pulled back the drapes and I was looking down from the second story of a residence. The street was new black top and it was lined with new round top trees. There was a sidewalk along the street and a sidewalk to the home I was in. There were homes on the other side of the street. It was night out and I seen no streetlights. The moon must have given the dimness.

I then turned around to explore the room and I seen a kid with blond hair of the age of about 17. It was his room because he was sitting to the left of the room. Before me straight ahead was a single door. He said to me "What am I doing in your dream?" I said, "This is no dream, I have been practicing spirit surfing." I then said, "I don't know how much time I have, so what's your name?" I shook his hand and he was real, because the flesh was real. I will disclose his name, even though I will be bombarded with inquiries from the whole country. He replied, "My name is Bates." It could be spelled differently.

I was then vacuumed through a tube and I could hear the "Whoosh." Everything went black, however I was fully conscious. I started to panic, feeling that I would not make it back to my world. So for that three or four seconds I prayed "Please help me get back to my room, please help me get back to my bed." I then arrived in my bed sitting upright and I opened my eyes to a sight that was not to be taken easily: <u>It was not my room!</u> The walls were hunter green, mine are white, and the furniture was different. However, my king size bed was in the same position as was my *body. I said with my voice shaking " This isn't my room." Then I was vacuumed through the tube again. I heard the same "Whoosh" and I then opened my eyes to the same bed same positioned body, but the correct room.*

Review
Later after other visits, I discovered the house I visited, was the home occupying the same space, but in the second world. The room with the kid was the one across the hall on the East side of the home. The floor plans are the same. Then the tube brought me across the hall to my bedroom, but it was the bedroom that overlapped mine in the second world. As for the kid named Bates, I called the phone book of my area. There were only two and they were related. I asked if any one died or was in a coma that fits the description given. The woman said "No." It could have been a relative of the previous first owner. Their last name was not Bates, however, families have lots of relatives with different last names.

JOURNEY TO ANOTHER WORLD THROUGH "THE SILVER CORD WORMHOLE"> A CREATURE EXTINCT IS ALIVE (LEVEL 1?)

It came to pass three days of enduring the method. I implemented the needed fatigue as night fell, with 80 proof liquor. Sleep befell me. My one body sleeps on Earth, while I walk in my other body on the other

Johnny Vincento

Earth: Which is the immortal world of Heaven. Neither Earth can see each other, but both occupy the same space.

I find myself standing in a home well decorated, about 20 miles distance. The name of the city in this world is River Forest. I was visiting the city that occupied the same space, so I don't know if they named it the same. The names of the streets were Franklin and Greenfield in this world, however it's the same space in that world. It was two homes to make one. The corner North East combined with the home of North. Walking through a room, there were a very many people around me, but not close to me. Before me was a fancy wooden coffee table, with a toy robot dog, made of plastic or aluminum. The color was gun blue and silver. It was about two feet long with four legs about a foot long each. It was about 8 inches wide. The only thing that was familiar with the toy, was its gun blue visor, covering the eyes. I seen those on toy robot dogs in this world, those toys are only about 6 inches high and 8 inches long. This toy was huge! I purposely went on this Journey to find information. I picked up this toy looking for a manufactures label. I turned it upside down and seen a blank square white label.

I walked on to cross the addition on the second floor that connects the two homes. The West side was all windows overlooking Franklin Ave. and the homes across the street. To my left were some dorm size rooms. One guy, about the age of 25, with black hair, was sitting in his room in the doorway on a black cloth chair. I asked, "All of you couldn't possibly be related, there are too many, do you all live in this home?" He said, "No one here is related, but we all live here and get along fine."

Stepping out of the overpass addition, I entered into the second home. Before me were two women with a creature that I have never seen: And the creature was alive! It was the color of dark olive green. I petted its head and its flesh felt like puddy. The living thing appeared to be a dinosaur. Sitting on rear two and standing on front two, its body measured at least two and a half feet high while upright – about two and a half feet wide – three feet long (without neck or tail) – the tail, about three feet long and pointed at end. While upright, its curved neck rose to my chest with its head easy to pet. The curved neck if straightened would measure about three feet long. Its head was about a foot or 11 inches wide, with some type of reptile lips. As for the description of the skin, it was one solid color and its texture was smooth like clay. The spine could be seen on its oval back. The snout of the animal was almost flat with the slightest of an arch, as was the top of its head, but rounding at the sides. The depth of the snout is unknown, for I was looking at it straight on. However, a guess of 4 to 6 inches would be round about. Its eyes were centered for front viewing, not like illustrations where the eyes are on the side of the head. As for the appearance of the eyes, they were *not like a reptiles, nor as a birds, or a cats, instead they were like a puppies and pure goodness emitted from its eyes and could be felt. For this reason, it had a spirit similar to a dogs and was above a reptile. <u>It resembled a Brontosaurus and the people had it as a pet</u>.*

With the quick sound of suction, the creature swallowed half of my right arm. I felt only pressure, no pain and I had no fear because pure goodness was within the living things eyes. The women started laughing and the one in the chair said "Don't worry, he always does that." The creature looked at me with its head tilted to my right and gave me a look as if saying "Are you going to stay?" Then the immortal creature let me pull my arm out and I returned to this world.

JOURNEY TO THE SECOND WORLD THROUGH "THE SILVER CORD WORMHOLE"> A TALK WITH IMMORTALS (CONFIRMED LEVEL 2)

In the year 2004, a Journey was taken like no other. I am a researcher and just because I do not understand, this does not constitute leaving the information out as did I when the original manuscript was copyrighted. The time of night on that weekday in this world was to cover the entire country, no matter what time zone. All other Journeys were similar: In the manner of weather and sky. If it was night out in the mortal world, it was night out in Heaven. If it was a beautiful summer day with clouds here, over there it was the same. Furthermore, all my "Teleportations" of arrival were the same except this one. The realm that was visited was as physical as all the other Physical Plains of existence. I give future researchers this left out beginning, because I do not know where I was<u>: Even though I was told exactly where I was.</u> This Journey above all, is what brought me into seeking the answers to time!

My body weak, I can feel that it is time. With my cats Snoopy by the window and Spirit by my side I prepare my questions in my mind. I slowly sip on 80 proof, the glass does its work. I pray to see all things unseen, rejoice! fright! excitement! fear! Creation opens as I close my eyes!

Immediately, I find myself resurrected in mid air in the sky! I was falling and in front of me was a nice cloud filled sky. After about three seconds I looked down toward the Earth. I could see the greenest grass, large trees and some apartment type rect-

angle buildings: Their numbers were three and positioned in a loose upside down "u" format. My vision was perfect. In this world my sight is like a digital image. As far as the falling, I was moving toward the ground, however, I was floating down. To elucidate, the atmosphere is thick but invisible. There was no "Gravitational Positive Force." It was like being quickly carried down at a comfortable speed. My lungs were not pushed up. <u>I was not carried, the atmosphere is different.</u> The altitude that I was at, I can only describe by an example. The largest branch spread tree looked as the size of an Eisenhower silver dollar being held six inches from my eyes. However, the vertical would be about 1/4 inch overlapping on top and bottom. I would estimate the tree top that I am referring to as being about 18 to 22 feet across. This example, was not a round diameter tree. It was the dimensions as stated so to say from my view point vertical. The horizontal so to say was about 15 feet across. All the trees had green leaves and with this largest tree, I could see the dark brown branches with my digital like new vision. This example is what I seen at my highest altitude minus three seconds I would estimate.

I found myself somewhere where it was daylight. Landing with little force and standing in front of two. Close to my right was a red brick apartment type building. One was a black haired girl about 5ft 5in and about 110lbs. She was a teen of about the age of 17 or 18 and she was wearing blue jeans. The other was a guy about the same age with an average build and taller. Standing in the green grass this is what was said:

"Where am I, is this place Heaven?"
"No, this is not Heaven."
"Then is this Hell?"
"No, this is not Hell either, this place is right under Heaven."
"Then am I in Purgatory?"
"There is no such thing as Purgatory: Come with us."

As I walked into a short hallway, in the front room was a naked blond girl sleeping on the floor. She appeared to be of the age of about 17 or 18 also. She was lying on her side and had long hair the length of her back.

"Do you want to be with her?"
"No, I am here to ask questions."

We went into the kitchen on the right and sat down at a round wooden table. I was on one side and the two on the opposite side.

"I am not from this world, I am from the mortal world."
The girl laughed saying: "If you're not from our world then how can you be here?"
"I came to visit, to ask some questions and when I leave, you will believe what I'm saying is true."
The guy said: "What questions do you have?"
"When one dies from the mortal world do they stay the same age forever in your world?"
"Yes."
"So you never age?"
"No we stay the same age."
"Then what happens when a new born baby dies in my world? Does the baby stay a baby forever in your world?"

Johnny Vincento

"No, that is different, the baby will age to 8 years old and stay at that age."
"Do you feel any pain?"
"No, there is no pain here."
"I seen that girls hair, does your hair grow?"

I then returned with rejoice of a successful Journey.

Update:

From my Journeys, the sun seems to rotate in the same position, equal in Heaven and Earth. For this reason, future unborn physics students must come to terms of other physical realms of existence. Living in "Past Time Realms" must be considered. Heaven is "Present Time Realm." This is why when the "Realm Boundary" veil is occasionally lifted, we can see the realm of Heaven.

I bring a final question to you of the 22nd Century. I have seeked to open time and the past can absolutely be visited: Your own past that is. How can the past be seen? It can be seen because it already happened. How can the future be seen if it didn't happen yet? I know the method to see the future, documented in "Father how long will I live?" It is written that past, present and future are all taking place at the same time. The answer to how the past can be seen may be the answer to how the future can be seen. When someone sees the past they are seeing the present. When someone sees the future they are seeing the present.

V

THE PURITY OF THE HEART

If you receive anything from my research receive this chapter and remember it for all your days. One reason you were born, is to purify your heart so you can become immortal in paradise.

Heaven is a misleading word, the term First or Second, world would be more appropriate for us. The other worlds can be paradise or they can be a living nightmare: That's when you go to level Three. You judge yourself, what is in your heart effects your surroundings. If your heart is pure, then you will live in paradise. If your heart is evil, it is written that you will go to a place in the other worlds that is within that worlds crust. That I have luckily not seen. This is called the "Resurrection of damnation." Then there is the darkness in the heart. Even the smallest amount of darkness can turn paradise into an eternal torment of terror.

Johnny Vincento

Update:

From the time I wrote this I've found that there is three levels of Heaven and the third is for all the mean jerks. Those not bad enough to go to hell.

JOURNEY

TO THE THIRD WORLD THROUGH "THE SILVER CORD WORMHOLE"> PARADISE TURNS TO TERROR

Obtaining the "Silver Cord," I found myself physically standing in a doorway of a home. I walked into the room and there was a small dark brown dog. I bent over to pet the dogs head. I was petting it and I thought to myself "What if this dog bites me?" As soon as I had that thought, the dog jumped up and attacked me. Biting my right arm, jumping up and biting over and over again. However, there was no pain only pressure. In the second world all your emotions are multiplied times ten. So my little bit of fear turned into absolute pure terror. So much so, that I flew back to my body in bed and sat up screaming!

JOURNEY TO THE THIRD WORLD THROUGH "THE SILVER CORD WORMHOLE"> PARADISE CAN BE PARADISE

Upset for two weeks about that terrible Journey, I was sad about my fate. I was sure that I didn't deserve such a future in the second world. Undergoing the fasting, the third night arrived. For the entire third day I prayed to my Father in Heaven. "When the dog bit me I was not injured." Sipping on 80 proof I began to become sleepy and this was night on a weekday. "Father in Heaven, I now know the answer: Take me to see the dog and I will have no fear. I am glad I went through the terror, it is better that I experience what awaits me for a few minutes, rather than an eternity."

Johnny Vincento

Closing my eyes as one candle flickers, I was immediately standing in the same place as before. I could see the dog and as before the rest of the room was empty of furniture and people. I walked into the room, which was about 25 feet long with a small hall and closed door on the opposite end of the room. As soon as I walked passed, the dog viciously mauled me: Worst than before. Jumping on my right arm and holding on me for seconds at a time. I was repeatedly bit on my right shoulder, right bicep, tricep and forearm. I could only feel pressure, no pain and no punctures. I kept on walking with no fear at all. I knew nothing could injure me in the second world if my heart is pure. At the end of the room the dog stopped, sat down and looked up at me. The dog had goodness in its eyes. I could feel the goodness and through the dogs eyes it communicated, "I thought you were still bad." I then bent over and petted its head. Then I returned to this world.

To get rid of ones darkness, pull it out by the roots once you recognize it.

Jesus said: "Verily I say unto you, whosoever shall not receive the Kingdom of God as a little child, he shall not enter therein." MK 10. 15-16

Jesus said: "Lay not up for yourselves treasures upon earth, where moth and rust doth corrupt, and where thieves break through and steal, for where your treasure is, there will your heart be also." MT 6. 19-21

Jesus said: "The light of the body is the eye: If therefore thine eye be single, thy whole body shall be full of light but if thine eye be evil, thy whole body shall be full of darkness. If therefore the light that is in thee be darkness, how great is that darkness!"
MT 6.22-23

It may be fitting for one to acquire a pet dog, when one receives the resurrection after death. For anyone with any darkness in their heart will be revealed by the dog.

NOTE: This was a great Journey for I turned (level 3) into (level 2) by having no fear and purifying my heart. The Gods shown me this lesson. Why else would I be in front of the same dog, same room, and same situation?

VI

Johnny Vincento

NATURAL LAWS OF INTERDIMENSIONAL TELEPORTATION

The Law of Secured Passage

There are natural laws not of this world, but between worlds. The Earth consisting of over lapping realms. Ours being the mortal, the others immortal. All occupying the same space, separated by dimensions and accessed through the resurrection at death or the "Teleportation" while alive. The other realms being just as physical as this one.

Attaining the "Teleportation" while one lives, secures the passage between worlds.

Doing this while alive takes away the risk of the Journey at death. This can be compared to a test run. Your spirit will already have secured the passage between dimensions by making your "Silver Cord" stronger. Then at death you will not go astray.

The Law of Duplication

The body is duplicated image for image. If you receive the resurrection, after death you will stay the same age for eternity: This being the real fountain of youth. The young staying young and the old staying old. The exception is for the too young. Refer to "A talk with immortals."

If one is without an eye or a limb, the resurrected body will be without an eye or a limb. It is important for one to understand this before one goes to the hospital for an amputation. This is a natural law that the world does not know about.

Did not Jesus appear to his disciples with nail wounds? I have seen the affects on the resurrected. In "Journey to my Father" the second woman's face was mangled. In "Children of Heaven" the one looked like he died of Chemotherapy; the other had a bullet hole in his head. The resurrection also duplicates the clothes one is wearing. The tattoo falling under the "Law of duplication" is unknown. However, this question can easily be answered.

The Law of Transfer

This law being two fold contains a timeline. The following are the issues:

1. The completion of the resurrection process at death takes up to 60 hours.

And

2. The compromising of the resurrection before the process is complete.

When one attains the resurrection during life or after death, ones clothes are duplicated. The woman in "Journey to my Father" was wearing a nightgown. This giving a clue of when she first started the resurrection process. The next clue being, she was gray, drained of blood and unable to function: Except for a shuffle.

Her resurrection process was compromised, most likely due to embalming. What happens when one is cremated? The spirit has no body to duplicate, because it was destroyed before the 60 hours. The same result as the angry spirits in "Accessing creation."

<u>After Jesus died</u>, was not his side pierced with a spear? This is damage during transfer. He appeared to his disciples:
"Then saith he to Thomas, reach hither thy finger, and
Behold my hand; and reach hither thy hand, and thrust
It into my side: and be not faithless, but believing."
John 20:27
<u>*Before 60 hours, there should be no autopsies, no organ donations and no embalming!*</u>

The Law of Function

This is the law of the miracle. Damage to the flesh is duplicated. However, damage to the nerves is not transferred over to the resurrected body.
Those with Arthritis, Cancer, Parkinson's, those paralyzed, etc. will be healed.

My dog died because of Cancer and could not use

her rear legs. When I seen her, she was walking on three. She doesn't know her fourth is good too.

Anyone who is paralyzed, do not take my word for the truth. How much will your spirit be healed if you take the Journey? You will walk while visiting another world.
"The Theory of Miracles," attempts to bring this law of function to our world.

This law transfers the mind and all its memories.

<u>Consequences</u>
The consequences of ignoring these four laws are the following:

<u>1. Your spirit being trapped in a non functional body.</u>
(The cause: The body was embalmed before the resurrection process was completed.)

<u>2. Your spirit being trapped in an amputated body.</u>
(The cause: The body unfortunately lost a limb during life.)

<u>3. Your spirit being without a body and floating around in pure spirit.</u>
(The cause: The body was cremated or destroyed completely at death or before 60 hours.)

<u>4. Your spirit is trapped in a mangled body.</u>
(The cause: The body was involved in severe damage at death or up to 60 hours after death.)

5. Your spirit is trapped "In the Middle" and dies.
(The cause: The spirit did not know the path to the other World.)

6. Your spirit enters your duplicated body and the resurrection process is a success.
(The cause: You attained the resurrection while you lived and your family protected your body for 60 hours.

***

In no way do I consider myself a lawgiver or a giver of laws. These laws are not written in stone. They are the compilation of the experiences that were seen. You may have your own conclusions by referring to my Journeys.

***

JOURNEY TO THE SECOND WORLD THROUGH "THE SILVER CORD WORMHOLE"> CHILDREN OF HEAVEN (LEVEL 2?)

In the year 2005 on the 15th day of May, I closed my eyes for the time was here. This was a Sunday at exactly 8:30 am.

I found myself "Teleported" in a moving vehicle. Driving the vehicle was a man I've never seen before. I was in the front passenger side of a possible 1960's black jeep. My Father in Heaven was in the back leaning between the front two seats. The man driving looked to be about the same age as my Father. He had black short hair and was Caucasian.

The Journey was a shock to me, because I lost my nerve at the last moment and had no one to wake me. In a total panic I started to yell "Where am I!" over and over again. I was banging both of my palms up and down on the metal black dash board. I must have hit that dash board ten times while thinking to myself "I want to know if this is real." I was slapping my face and thighs with both hands while still yelling "Where am I!" The driver of the vehicle said to my Father, "What's the matter with him?" My Father answered "He doesn't believe this is real." In that realm telepathic thoughts can be picked up.

While in the vehicle I could see a nice country dirt

curving road with short grass fields, trees, and a nice sky. I could hear the road and any gravel that the tires ran over.

As for the description of the vehicle, it is as follows: I thought it was from the 1960's, it had a large steering wheel, left and right lever vents to where the heat controls are. There was no radio however a slot was empty to where it should be. Between the seat on the floor board was no gear shift, it may have been on the steering column. The dash board was all sheet metal, no plastic overlay and there were no emblems. The jeep had a roof and doors. The inside panel of my door felt like it was covered in plastic, I was banging on it. The black hood was not square. It sloped in toward the front. Larger by the one piece curved windshield and sloped at the nose. As for the means of moving, It was absolutely silent, not a sound except of the wheels moving over the road. I did not feel any transmission shift nor see the driver shift.

I yelled "Stop the vehicle right now!" I knew I could not be killed or injured in that world so I was going to jump out of the moving vehicle. While opening the door I attempted to jump out. The driver put his right arm over me at my chest. I could not move but remembered my research and said to myself "No one can over power me in this world." With my right hand I pried his fingers and right wrist all the way back. The Jeep stopped and I got out.

I found myself standing in a field with a large tree to my left. I felt my legs, arms and put both hands on my face. I thought to myself, "I'm here again." The worlds are exactly equal physical wise and conscious wise. That is how you will always be able to tell the difference between a dream and the "Teleportation."

*About 35ft ahead of me was a single story home with a **wooden** outdoor deck. The deck had a balcony around it. There were a very many people as if at a get together. I walked into the gathering and seen a disturbing sight. A young man was sitting in his shorts with absolutely no hair, not even his eyebrows. I figured later that he died of Chemotherapy. I then passed a man with a bullet hole in the back right side of his head. I could see the powder mark. All the kids were on the deck, which was about two steps off the ground. The color of the deck was unfinished pine looking wood. The deck was attached to the left side of the home.*

As I turned left around the deck to the front of the house, two kids approached me.
One was black haired about the age of ten. The other was about 13yrs old with blond hair and taller. This is what happened:

"Do you want to see something?"
"Sure what?"

They pointed down between them and there was a

round ball, the size of a volleyball. It was the color of liquid water. I could see the white pavement through the water. The first 10% was like the color of pale flesh, however, the correct color is not a color at all. I can only describe it as looking through hot light soaped water, into a white porcelain bath tub. Now this ball was a spirit and moved as the kids moved. I said:

"Yeah he's really cute."

I could not see the face of the spirit. I was looking down on it as it hovered a bit off the pavement. And even when I did see the part where the face should be, this being later before I left, it did not have a face! It was a round spot which was the color that I am trying to describe. It resembled a circle within the sphere of liquid water. I got the feeling that this ball was the spirit of a child, possibly one not even born yet.

I then seen my Father in Heaven walking onto the front of the driveway. My spirit rejoiced while I walked up to him. He took my left hand with his right and we walked to the home. The rejoice was short lived. At the front of the home I stopped and jerked his hand back. I said:

"Are you my Father in Heaven?"

Telepathically, within my head he said:

"What, you don't know who I am?"

I panicked again and started yelling to the people around me:

"Where am I, is this place Heaven?"

No one would talk to me as if they were not suppose to give me any information. However, the blond 13 year old kid leaned down toward my ear, from standing on the first stair of the balcony and said very quietly:

"Everything you are saying is true."

Then the two started talking among themselves about if I was one of them. While they talked I looked over the many people which was over fourteen. I recall the 10 year old very specifically within their conversation saying:

"Maybe he's not one of us."

After their debate, the 10 year old said to me very straight out:

"Are you a nighty?"
"I don't know, what's a nighty?"
"They usually come to visit at night time and then they are picked up. Are you going to get picked up?"
"I hope so, how about you, are you ever going to get picked up?"
"No ------- I'm never going to get picked up." (And he said that with disappointment)

I then returned and opened my eyes to my world and the weather was the same.

Johnny Vincento

Recap:
I drove electric machines before, and they make some sound. If you refer to my experience with the man with no face: This being documented in "The Theory of matter." You will see that in the world of Heaven the mind can move objects. The research after this Journey shows how the vehicle was able to move with no power source. I would have left the beginning of this Journey out, however further research supplied the answer to how the vehicle moved. Also, I did not want to scare anyone because of the panicking. Learn from my mistake. If you lose your nerve about taking the Journey, call a friend to wake you after 4 minutes 45 seconds. Once the third day of fasting hits, the formula is activated whether you want to take the Journey or not. So compose yourself, be of strong courage and most of all, have the security in your mind of your friend waking you.

THE THEORY OF MATTER

It is my theory that all things in the overlapping immortal Earths are made of the same elements. In many ways this is the theory of creation. Being a "Physicist Pioneer," it is difficult to comprehend what is impossible in our world. However, a natural law of elements, subject to thoughts, in those physical worlds.

So let's say that in the other immortal worlds the atmospheres are a sea of invisible building blocks of creation. If you wish for a loaf of bread, it appears. If you wish to turn water into wine, it's done. If you wish for a stainless steel knife, it appears out of thin air. You see, "The turning of water into wine" was considered a miracle. However, Jesus was a doorway between worlds. Thus, showing us a natural law of physics which is common in any of the other physical immortal worlds.

"For verily I say unto you, that whosoever shall say unto this mountain, be thou removed, and be cast into the sea; he shall have whatsoever he saith."
"Therefore I say unto you, what things so ever ye desire, when ye pray, believe that ye receive them, and ye shall have them." MK 11. 23-24

How can this be, even if the building blocks of creation are in the sea of the atmosphere. Proof alone is the resurrection of the flesh, image for image. If one is without a foot, one enters the other overlapping world without a foot. If one enters into the resurrection without an eye, then your image of your flesh will be without an eye.

Jesus said: "If the flesh came into being because of spirit, that is a marvel, but if spirit came into being because of the body, that is a marvel of marvels." Gospel of Thomas 29

If you read "Journey to my Father," I made a stainless steel knife appear out of thin air, just by wishing for it. I could feel the metal and it was solid.

We were born with a brain that we only use a cer-

tain percentage of. My theory is that the part we don't use is used in the immortal world. Without a doubt it is a transmitter which manipulates your surroundings: Physically and emotionally. For all emotions are times ten over there. The mind creates physical things out of the building blocks that are in the sea of the air.

The scientific elite calls this "The String theory." That all things are made of the same element called "Strings." They have a theory that can't be proved on Earth: Except for on paper. Good news for them though, because their theory hit exactly on the mark concerning the other worlds.

To the scientific community I say: Follow me either now, or a hundred years from now. History can't stop what I brought to the world.

To all Physics students. Try to prove my research wrong or correct with your equations. It will make a great Term Paper.

We are made of particles and particles can be in multiple places at the same time. Thus, in "Father how long will I live" I was in three places at once. One on this Earth sleeping - One in front of my Father in another world - One going back in time to when I was a kid looking at the indian toy. See, three places at once. Furthuremore, in the Journey of the three crosses mentioned later, I also put my fist through a metal container. Then pulled my arm out and the metal reformed itself to be perfect again. My arm had no problems. Now this was one of many worlds and not all worlds are this way. In "Journey to my Father" I was banging my head against the wall and felt no pain, only pressure. That wall was solid and my head didn't go into the wall.

Update:

Those of you who are students of New Laws of Physics: This being those of the unborn. I will add information that was not included. All my Journeys are not documented in this compilation. Only the ones giving the most information. This numbering the ten included. However, there is a Journey that will provide you with an insight to the other worlds physics.

To briefly explain, I was resurrected in front of the warehouse that overlapped my families business in the other realm. I patted myself and felt my thighs and arms. Both worlds are equal, one cannot even tell the difference physical wise or conscious wise. I went into the front office area and then into the kitchen dining area. There were no lights, only numerous candles. I met a guy who had his face blacked out. The rest of him I could see. He was solid and I could see his clothes. We were sitting at a rectangular wooden table. He was facing North, I to the South. To my right, touching the wall, was a stack of white standard looking paper. The size was that of a stack of about 250 sheet block, like you see at the store. He said to me "Watch this" and he pointed his right index finger at the stack and made a single sheet of paper hover about a foot off the table. Then he quickly moved his finger to his right and the single sheet of paper shot across the room straight like a dart. I then heard it hit the wall with force. I said "That's neat do it again." He then did the exact same thing.

VII

INVESTIGATION INTO THE RESURRECTION OF JESUS

Intended is to prove the following:
1. Jesus did not walk out of his tomb.
2. Jesus received the resurrection first in the second world.
3. Jesus performed "reverse resurrection" to visit this world.
4. Jesus returned to the second world.

Before one begins this investigation, one must take the witnesses accounts as factual. Their stories are irrational and that makes them more valid. All four Gospels are describing the same event. Thus, the more detailed description must be taken over the vague one. This is not pick and choose, because all four are the same account. Now the complete story begins:

FIRST EVENT

"And when she had thus said, she turned herself back and saw Jesus *<u>AND KNEW NOT THAT IT WAS JESUS,</u>* Jesus saith unto her, woman, why weepest thou? Whom seekest thou? She, supposing him to be the gardener, saith unto him, sir, if thou have borne him hence, tell me where thou hast laid him, and I will take him away. Jesus saith unto her, Mary. She turned herself, and saith unto him, Rab-boni; which is to say, Master. Jesus saith unto her, touch me not; for I am not yet ascended to my Father; but go to my brethen, and say unto them, I ascend unto my Father, and your Father, and to my God, and your God." John 20: 14-17

SECOND EVENT

"And it came to pass, that while they communed together and reasoned, Jesus himself drew near, and went with them. But their eyes were holden that *<u>THEY SHOULD NOT KNOW HIM."</u>*
Luke 24: 15-16

"And it came to pass, as he sat at meat with them, he took bread, and blessed it, and brake, and gave it to them. And their eyes were opened, and they knew him; and he vanished out of their sight." Luke 24: 30-31

THIRD EVENT

"Then the same day at evening being the first day of the week, when the doors were shut where the disciples were assembled for fear of the Jews, came Jesus and stood in the midst, and saith unto them, peace be unto you. And when he had so said, he shewed unto them his hands and his side. Then were the disciples glad, when they saw the Lord." John 20: 19-20

FOURTH EVENT

"And after eight days again his disciples were within, and Thomas with them: Then came Jesus, the doors being shut, and stood in the midst, and said, peace be unto you. Then saith he to Thomas, reach hither thy finger, and behold my hands; and reach hither thy hand, and thrust it into my side: And be not faithless, but believing. And Thomas answered and said unto him, my Lord and my God. Jesus saith unto him, Thomas, because thou hast seen me, thou hast believed: Blessed are they that have not seen, and yet have believed." John 20: 26-29

FIFTH EVENT
"And it came to pass, while he blessed them, he was parted from them, and carried up into Heaven." Luke 24: 51

SIXTH EVENT
"But when the morning was now come, Jesus stood on the shore; *BUT THE DISCIPLES KNEW NOT THAT IT WAS JESUS.*" John 21: 4
"Jesus saith unto them, come and dine. *AND NONE OF THE DISCIPLES DURST ASK HIM, WHO ART THOU?*" John 21: 12

THE ISSUE OF THE TIMELINE
Jesus died on Friday at the ninth hour. In the first event, Mary Magdalene encounters a man, not in the image of Jesus. Even so, this messenger spoke as Jesus and for Jesus.

The messenger said, "Touch me not; for I am not yet ascended to my Father." John 20: 17

This encounter was Sunday, "And very early in the morning the first day of the week, they came unto the sepulcher at the rising of the sun." MK16:12

In those times *Morning* was 3am to 6am, the *Ninth hour* of the day was 12pm to 3pm and *Evening* was 6pm to 9pm
The time line is as follows for the resurrection of Jesus

****
****

Friday between <u>12pm</u> to *3pm* to Saturday
12pm to <u>3pm</u> was *21* to <u>27</u> hours
Saturday between <u>12pm</u> to *3pm* to Sunday
6pm to <u>9pm</u> makes *27* to <u>33</u> hours total
The resurrection at minimum took 48 hours
The resurrection at maximum took 60 hours

****
****

The messenger seems to verify that Jesus's resurrection was not complete yet. When Jesus appeared for the first time, <u>*as himself in his own image,*</u> it was "The same day at evening being the first day of the week." John 20: 19
And this would make only 60 hours maximum. So it seems Jesus was referring to the days of the week and not the hours in a day when he said, he would rise on the third day.

<u>THE ISSUE OF THE TOMB</u>

Jesus could not have walked out of his tomb. I understand that his body might have changed appearance, due to approx. 40 some hours of being dead. This being the first encounter with the messenger at approx. 6am on Sunday. It is written, "He appeared in another form." Mark 16: 12. Witness accounts tell us it was not his body.

They did not recognize him. One cannot be unrecognizable in the morning and recognizable in the evening. He was recognized in his own body that evening. Furthermore, the <u>mortal</u> body cannot appear out of thin air in the flesh through dimensions. One can rematerialize their second body in the other realm and go exploring. Or one can visit our world from that realm in their immortal body. As did Jesus and that solid looking ghost who visited me on July 7th 2008. However, Jesus may have been half angel. Read update at the bottom of this subject.

THE ISSUE OF THE RESURRECTION

The world doesn't understand, Jesus was resurrected in the second world. While alive he suffered nail wounds. The wounds were duplicated to his resurrected body: *"Law of Duplication."* While dead he suffered a spear wound. The wound compromised his resurrection process, transferring the damage to his resurrected body: *"Law of transfer."*

THE ISSUE OF REVERSE RESURRECTION

There is valid descriptive evidence, of Jesus's messenger, performing reverse resurrection. "And he vanished out of their sight." Luke 24: 31. This is how it is when one visits through dimensions. Just as I have appeared out of thin air, in the flesh and disappeared. However, the messenger and Jesus did it in reverse. The description of Jesus's two visits are very accurate. Jesus visited twice, once on Sunday and the second time eight days later. Both witness accounts say that Jesus appeared in a closed room with the doors shut, in the flesh. One cannot get a more accurate description of an interdimensional visit through the resurrection. I have attained the resurrection and that is exactly what happens.

THE ISSUE OF JESUS RETURNING TO THE SECOND WORLD

After the second visit, witnesses say that Jesus was *"Carried up into Heaven."* **Luke 24: 51**

This is very interesting, because in "Children of Heaven", the one child describes what he sees:

"Are you a nighty?"
"I don't know, what's a nighty?"
"They usually come to visit at night time *and then they are picked up.* Are you going to get picked up?"
"I hope so, how about you, are you ever going to get picked up?"
"No ------- I'm never going to get picked up."

Update:

In the year 2018, after looking closely at the Shroud of Turin. A spark entered my mind. "The transfiguration of Jesus" as written in the scriptures. He transformed into a Heavenly being shining the brightest white light while standing with Moses and Elijah. After Jesus was done, he told the three of his disciples not to tell anyone what they saw until after his resurrection.

So it may be that while alive he was able to walk in and out of Heaven.

What non-believers don't want you to know about the

shroud is that the image was made by light. If you look at Jesus' fingers, you can see the bones under the skin. And if you look at his closed mouth, you'll be able to see his teeth through the skin like an x-ray: <u>Like the transfiguration!</u> Furthermore, there are no drag marks on the cloth if the body was stolen. And what they don't want you to know as well, is the blood type is AB and there are two period coins over his eyes that you can read.

JOURNEY TO ANOTHER WORLD THROUGH "THE SILVER CORD WORMHOLE"> FATHER HOW LONG WILL I LIVE?

In the year 2006, on the 26th day, of the third month, I was two full days fasting past. This was a Sunday, in the early dim hours. I did not expect the Journey so early, but it goes to say I was very weak.

I found myself in a courtyard surrounded by a wooden fence. The fence was four inch boards standing side by side and each was rounded at the top. The sky was just starting to become a dim white. I yelled out "Father where are you?" *and I seen him very quickly. He appeared in the image of my own father on Earth and was in the flesh for I shook his hand. This is what was said:*

"Father it is good to see you again. I took great effort to see you, but our time is short together. I came to ask you some things."
"What do you want to ask me?"
"I want to see the future."
"Not everyone has the ability to see the future."
"I want you to show me the future."
"Well let's see if you possess the gift."
"I want you to concentrate on where I am pointing and think of seeing the future."
"You mean between those two trees over there?"
"Yes, do you see any visions of the future?"
"No."
"Let's see if you can see the past. Concentrate on where I am pointing and think of Indians, do you see any visions?"
"No."
"Then you do not possess the ability."
"Father how long will I live?"
"To seventeen."
"I am much older than seventeen."
"Everyone has an inner age."
"Like an emotional age?"
"Yes."
"But I want to know how much longer I will live?"
"Don't worry, you have many numbers."

Then we walked into a department store type setting. The courtyard was connected to the building. There was a great amount of clothes. I walked to a room that had no door, and seen a woman washing her hands in a sink of about 60's in age. I said:

"Excuse me, how old are you?" and she replied slowly "Fifty thousand years."

Then I approached an old lady of about the age of late seventies in the main room. She was five feet or five feet and a couple of inches tall. She had gray hair and if there was a line up of photographs, with her among them, I would be able to pick her out. I asked this lady:

> "Excuse me maam, when were you born?"
> "I just arrived here yesterday at about 4:30."
> "So you died yesterday?"

Then a light skinned African American speaking woman, who

was standing to our left, said:

> "You are not supposed to disclose that information to people."

She was dressed in a security outfit dark blue in color. Her hat did not have roundness to it. Instead, it was as the shape of an octagon or a six sided hat about the top. And there were no insignias upon her uniform. She was about 5ft 7, weighing about 125lbs. After the security guard spoke to me, I figured that this lady and all the many people amidst the store were not being told that they were even dead. For the old lady seemed to not even realize it. I then yelled out:

> "Father where are you?"

I seen him about ten feet away, went up to him and

put my left hand on his right back shoulder and said:

"Get me back, I want to go back right now!"

We walked into the double glass door enclosure that department stores have. Two glass doors, then a small enclosure, then two more glass doors to exit. We were in the enclosure when I was grabbed from behind by the security guard. She did not want me to leave for reasons unknown. I said quickly:

"Father, help me!"

I had no problem breaking her hold on me from behind and giving her a half toss push. I then saw a young guy of about 24 approaching the enclosure from inside the store. I could feel his concern for my well being. I then looked at the woman in front of me and thought:

"I have no fear of you; you can't hurt me in this world."

This all happened in seconds and after I thought that she put her head back in surprise as if she just understood what I thought. Then a force came between us and this force was invisible. It communicated to me through thought that if I choose to fight her, me and her are equal in physical strength. I put my fist up, to ready for her next advance and it was made clear to me that if I punch through the force barrier, my punch would only have half of the impact power. This was not a lie either, because I felt half my strength leave me as soon as I raised my fist. I was not going to back down even with half strength. Then the woman tried to scare me and her true meanness and darkness shown.

She put both hands up by the side of her head and made cat paw fingers. And then her eyes were dilated open all the way black, while she moved her head forward making a cat like hiss. At this point I was surprised, but I tell you, I learned my lesson in "Paradise turns to terror," I had absolutely no fear and I was not going to back down, I was not going to let this mean spirit control me or torment me. And again I say, all this happened in seconds. I then said:

"Father, help me."

Then he helped me by returning me to this world.

Review

As you read the conversation with my Father in Heaven, that is how fast it was, so to say, in real time. I did not have time to concentrate on seeing the future. The only thought I had when he was pointing, was that I want to see the future. Concerning the attempt to see the past, I did see the past. It was however, only for a quarter of a second. I seen my body, lying on my right side on a floor, I could see my front thighs and I was wearing blue sweat pant type material. I was looking at an Indian toy that I received from an island in Lake Michigan when I was about 5 or six years old. The toy was a painted wood carved totem pole. This toy is what I was thinking about while he was pointing. And after some years I understand why he said "and think of Indians." One must think of something very specific to see the past or the future. That is if one has the ability to see the future. As far as my vision, it was complete, as if I was there. Everything with my Father happened so fast and the vision was not even a second long. If I could have concentrated, I would have been able to see more: At least for the past. As far as the future goes, I did not concentrate.

At first I did not want to put in what I had experienced with the security guard or the lady that said she was 50,000 years old. Too bizarre. I put it in, because this work discloses many things, good and bad, explainable and unexplainable, the deciphered and the undeciphered. However, "This work" does not disclose everything I have seen and experienced.

Update:

I thought I was going to die in 2017 but it makes sense now. Because, I was Baptized on December 4th, 2016 and I was REBORN so the old me DIED in 2017. Before I went under the water I said "OK this is it, I repent for all my sins, I take Jesus Christ as my Lord and Savior. The karma is over, there is no more payback. I also repent for any sins I may have done in a previous life if I even had one." Then I went under the water. And this is how I removed my karma payback. That was the greatest day of my life and I never had any payback Karma again. My Pastor was walking with me after and I felt like I was walking with Jesus himself.

VIII

THE SAYINGS OF THE SPIRITNAUT JOHN VINCENTO

1. Rejoice to the Keepers who hold my carvings with a pure heart. For I will be asking my Father in Heaven about them and I will ask him to give the Keepers blessings.

2. Rejoice to the one who deciphers what is still undeciphered in the Gospel of Thomas.

3. Greatly disturbed will be the one who deciphers the Gospel of Thomas number Eighty four. If you can, leave this saying be. You will not like what you find.

4. Rejoice to the one who makes gold in Heaven and plans to make gold on Earth.

5. Rejoice to the one who sees a creature that used to roam the Earth. That one will truly be happy.

6. Woe to the evil one who abuses children or kills animals for fun. You will obtain the "resurrection of damnation" in the "Third World." The location – cavities within the Earths crust.

7. Rejoice to one who wants to reign in Hell and takes the Journey. Then comes back and becomes a Saint.

8. Rejoice to the prisoner who can leave his cell any time he plans to.

9. Rejoice to the one who was born blind and took the Journey and sees for the first time.

10. Rejoice to the father who seen and held his own child, even though that child died years before.

11. And Woe to the Evil One who killed that child. For that one will live within the Earths crust in terror times ten.

12. Rejoice to the one who becomes a doorway between Heaven and Earth and walks again in both realms.

13. Rejoice to the one who travels through time during the Journey.

14. Rejoice to the one who finds The Spirit of Truth and has questions ready to ask about the future.

15. Rejoice to the one who gains cures for our world by traveling to another realm for knowledge.

16. Rejoice to the one who asks his Father in Heaven about Intelligent life on other planets. Along with, how to travel the universe to get there?

17. Rejoice to the one who starts experiments on the "D.S.V. machine and knows not all homes are overlapped.

18. Rejoice to the one who asks his Father in Heaven about the Creation of the Universe.

19. Woe to the one who mocks my teaching. For it is better that you have never received it and are ignorant, rather than have received and rejected it. For you will be rejected from Heaven.

20. Rejoice to the Greedy One who makes gold in Heaven. You will do more for the Mystic Art than the pious.

21. Rejoice to the one who seeks the lottery numbers in Heaven. You will be guaranteed the "Tele-

portation."

22. Rejoice to the aviation engineer. For there are different laws of physics in that realm of existence. That world does not contain a positive gravitational force.

23. Rejoice to the one who goes to Heaven to find out if the tattoo is duplicated.

24. Woe to the one with a tattoo of Satan if it is so.

25. Rejoice to the one who plans on traveling to Heaven to obtain the secret of Anti Gravity.

26. Rejoice to the one who knows the laws of the resurrection and chooses to die, rather than being cut apart into pieces. For that one will be whole in Heaven.

27. Rejoice to the one who already walked in Heaven and knows he will not die.

28. Rejoice to the one who knows where and when a catastrophe will hit. That one can save himself and his family. For no others will believe him.

29. Rejoice to the one who seen and held his own dog. Even though the dog died a year before.

30. Rejoice to the one who cannot see without

glasses. For that one will have perfect vision in Heaven.

31. Great is the one who does not kill a mouse with a trap. Instead, captures it and sets it by the river with some food.

32. Great is the one who does not kill an insect that is bothering no one. Instead, captures it and sets it outside.

33. Great is the one who reached those two. For your heart is becoming purified.

34. Comfort to the child with cancer who reads my words. You were a spirit in Heaven before you were born. Be not afraid, for you will not die. You had to be born so your spirit could have a body. And now that you have a body on Earth, your spirit will have a body to duplicate in Heaven. For you will now be whole.

NOSTRADAMUS: ONLY A MYSTIC CAN UNDERSTAND A MYSTIC

"Sitting alone at night in secret study,
At rest on a stool of brass? A slight
flame leaves the solitude,
And prospers that which should not be accepted emptily."
Century one – one Nostradamus

"The divine verb will give to the substance,
That which contains the Heaven and the earth,
occult gold to the mystic deed: Body,
soul and mind are all powerful.
All is beneath his feet, like the seat of the Heavens."
Century three – two Nostradamus

"Only those divinely inspired can predict particular
things in a prophetic spirit."
Letter from Nostradamus to his son Cesar

"Moreover when ye fast, be not as
the hypocrites, of sad countenance: for they
disfigure their faces, that they may appear
unto men to fast. Verily I say to you.
They have their reward."

"But thou, when thou fastest, anoint thine
head, and wash thy face;"

"That thou appear not unto men to fast, but
unto thy Father which is in secret:
and thy Father which seeth in secret: shall reward
thee openly."
MT 6:16-18

It is not opinion that the mystic of the 16 century read the Gospel of Matthew. In a letter to his son, he writes, "Give not what is holy unto dogs, nor cast your pearls before swine, lest they trample them under their feet an turn and rend you." This Gospel is before the King James Version and can be compared to MT 7: 6. The large book contains the chapters of Matthew 6 and 7 on the same page.

There is no respect for one who has the ability to see through time and plays word games. I have no opinion on the interpretations of his game playing. This chapter was included to show light on the method he took. And indeed, without a doubt, not using an opinion, he did have access to Heaven: Only a Mystic can understand a Mystic.

"Sitting alone at night in secret study." ***** Matthew chapter six describes, going into your chamber or closet and shutting the door behind you and seeing your Father in Heaven in secret. And as you can see, fasting. Just as I, he considered this experience to be research.

"At rest on a stool of brass?" ***** Notice the question mark. The word rest is a very special tool in perfecting the mystic art. For example, the day of rest is the Sabbath and this is the first day of the week, being Sunday. And Sunday is the only day that a mystic can visit his Father in Heaven. I have attained the "Teleportation" on different days and nights. However, the only time I was ever able to see my Father was on a Sunday. This word has a double meaning also. Rest is not only Sunday, but sleep as well. One can only open the door to the second world by going to sleep combined with great fatigue and fasting. Nostradamus's son Cesar wrote that his father only slept around three to four hours a day. As for the stool of brass. That translation has a couple of different interpretations, "stool of brass" being the favorite. The properties of brass contacting the body, while one is a doorway, may provide beneficial effects: This is unknown research. And I can add one other interesting observation. When one is sitting upright as described in my Journeys, "Accessing Creation," and "The Birth of the Spiritnaut," the individual has much greater abilities.

"A slight flame leaves the solitude," ***** This seems to be an experience I had in "Accessing Creation." It is most likely that he went to sleep, attained the "Wormhole" and then woke up in an upright position. Before him was his room for a moment, then the doorway opened to the world he was just at, for he was the door. It is an amazing thing, to be looking into your dark room and suddenly the light of the other world breaks that solitude.

"And prospers that which should not be accepted emptily." ***** This speaks for itself.

"The divine verb will give to the substance," ***** As I have said, the words in my book can be activated by the individual. The source of the words I have attained, are from Jesus Christ. The meaning is, The words of the Gospels will give you the ability.

"That which contains the Heaven and the earth," ***** Now add the rest of the sentence and you have, The words of the Gospels will give you the ability to travel to Heaven from Earth. The original text translation for this is, **"That which contains the sky and the land,"** The meaning is Heaven and Earth. I know this, because he is doing the mystic art that I undergo in a very similar way. The difference is that he was born with the ability to see the future, as I found out through my own research in "Father how long will I live?"

"Occult Gold to the mystic deed:" ***** An ordinary man seeks material wealth, but a mystic seeks occult wealth. I wish I had someone to go to for this wealth, instead I have become the pioneer. If someone informed me on how to reach certain goals, such as having control over time while visiting Heaven, that would be very valuable to me! Note: The mention of gold in "Sayings of the Mystic John Vincento," have nothing to do with this verse.

"Body, soul and mind are all powerful." ***** My research discovered that the body is duplicated in Heaven and the spirit can travel to the second Earth and rematerialize that body. This being accomplished through the different laws of physics: Specifically, "The Theory of Matter." The mind travels with the spirit and keeps its memories and personality. From what I experienced in "The

Birth of the Spiritnaut," the mind does not even need a body.

"All is beneath his feet, like the seat of the Heavens."
***** As Jesus said, "Know what is in front of your face, and what is hidden from you will be disclosed to you. For there is nothing hidden that will not be revealed."
Gospel of Thomas 5

The revelation of this verse came to me in the morning while waking in a Winter month:
"The wand in hand placed between the branches, He shapes the edge and foot," Century one – two
This can be while he was a doorway sitting in his room, after he came back from his second body, or while he was in Heaven in his second body. Ones clothes are duplicated with the body, so his wand may have been also. Concerning the method of seeing the future, one must point and concentrate in a specific area in the empty air. Also, the empty air may very well need to be outlined. It may not be a coincidence that my Father in Heaven pointed to where he did.

The following is taken from "Father how long will I live?"

"I want to see the future."
"Not everyone has the ability to see the future."
"I want you to show me the future."
"Well let's see if you possess the gift."
"I want you to concentrate on where I am pointing and think of seeing the future."
"You mean between those two trees over there?"
"Yes, do you see any visions of the future?"
"No."

Johnny Vincento

The wand is for the same purpose as was the pointed finger. The remainder of the verse, compare to my experience and make your own conclusions. The trees outlined the sides and the top, the fence outlined the bottom. This made a frame around an empty sky *"He shapes the edge and foot."*

In closing, prophets have to be born with the gift of seeing the future. However, I believe that there is one opportunity that has not been attempted in my research. That is contacting the Spirit of Truth. As for seeing the past, that had been achieved for a quarter of a second. One needs to concentrate and not be rushed. It needs *to be tried again*, because now it is known that it is possible, for ones past life time at least. There will be those who are truly gifted *at birth.* Those should be *blessed in Heaven with the ability* to carry on my teachings and perfect their skills.

Update:

"The moon in the middle of the night over the high mountain. The young sage alone with his mind sees it. His disciples invite him to become immortal. his eyes in the middle, his hands on his breast. His body in the light." 4-Q31

Over the high mountain is from revelation 21:10 "So he took me in the spirit to a great, high mountain…".

"Second to the last of the sirname of the prophet. Will take Diana's day as his day of silent rest. He will travel far and wide frantically to deliver a great people from subjection."

2-Q28

Diana is the God of the Moon. My birth sirname is from Poland. It is in two parts. The second to the last is "LAH" which means God of the Moon.

THE ALPHA AND THE OMEGA

Millions of people pray to God and they have no understanding of the greater scope of existence. It would be much more beneficial to pray to ones Father in Heaven and to understand the words that I scribe.

The masses need a Shepard because they grieve greatly. This is because they have no knowledge of their own spirit. An absolute gigantic amount of people throughout the world believe only in the here and the now. You are born and then you die and there is nothing else. This is because there are very few Shepard's. My intensions are to heal the spirits that are crying out. When one feels their own mortality is going to end, that's when people can actually feel that they have a spirit. This is unfortunately due to one's spirit crying within the body.

So now a "Unique One" arrives, to heal who? Certainly not the present age, the Enlightenment of the *things in my research are not even on the horizon for the people alive in my age. I send these books into the world as seeds: To help a few now and to blossom for the unborn of the 22nd Century.*

One will no longer have to say "The Alpha and the Omega." Those who understand and for the bold who wish to see. It will be more fitting to say "The Alpha and the Omega and the Alpha."

So now I will tell you about the cycle of things:

1. You were created out of water and your Father in Heaven also gave you consciousness.

2. Your Father in Heaven is also in pure spirit form at this time.

3. You are also given a companion spirit called "The Holy Ghost." The explanation of the Holy Ghost is a complicated one. However, all three of you may be connected by invisible cords. Now this is a theory which is not written in stone, but explains everything very well.

4. You are sent to Earth to attain a body. You have to be born on Earth so your spirit will have a body to duplicate in Heaven. Jesus said, "How miserable is the body which depends on a body and how miserable is the soul that depends on these two." Gospel of Thomas.

5. At the same time you are being born: Through invisible cords your Father in Heavens spirit receives a body: The body of your Dad at his present age. This is how every-

one's Father in Heaven always looks like their own Dad on Earth. As long as your Dad is alive on Earth, your Father in Heaven will age right along with him. If this was not the case, my Father in Heaven would not have been in his Sixties, just like my Father on Earth: He would have stayed 26 years old. The age he was when I was born. So this is why you have two Fathers.

6. Your Holy Ghost does have the ability to see the future and your surroundings for possible prosperity. This is why you have thoughts not your own. When one pays no attention to them, that's when one gets killed or almost killed. That was your Holy Ghost warning you before hand. Everyone has their own Holy Ghost. When you met someone because some thought said, "Go into that place." That was the Holy Ghost. When you meet someone though: Whose Holy Ghost is responsible? You may have been a learning tool for someone else's Holy Ghost. This may be the reason you feel great loss when you lose a relationship. The question of all time is, why do bad things happen to good people? Jesus answered his disciples in the New Testament. He said sometimes bad things just happen. This was a reference to when an aqua duct fell and killed some workers. Putting aside that, the main reason is "Dark Angels." They are rogue spirits and

attach themselves onto the invisible cords. Then influence people to do terrible things. There are also positive views that come out of bad events. Just as described in "Children of Cancer." One should look at the all encompassing view of ones existence. People are only on Earth for a very short time, Heaven is eternal. I seen that creature extinct, that gives a good idea of eternal. I was attacked on Easter last year. *On a personal level,* that turned out to be the greatest thing that happened to me in fifteen years. The chain reaction brought me prosperity of health. Look at the terrible things that happened to Jesus Christ. It was also the greatest thing to happen to the mortal world.

7. Concerning the "Cords Theory," all children will be going to Heaven, their cords are strong. When one gets older one's cords break or become weak. This is the reason it is written all over the Gospels, that one must seek out ones Father in Heaven while one is still alive, otherwise you may not be able to find him when you die.

Know the greater scope of existence and that knowledge will benefit you.

IX

LETTER TO THE UNBORN OF THE 22nd CENTURY: THE REALM BOUNDARY AND THE D.S.V. MACHINE

There are those who do not understand of what cannot be seen. Never arising past the eyes of mortal existence. I risked my life so to enlighten the world. Through my risk of crossing into the unknown, I returned every time, so all will know that the Journey is safe to travel.

UPDATE: This info is describing only the immediate closest overlapping world.

The "Realm Boundary," is a dimensional space which veils our physical world from the other physical world. The scientific few, describe this as a wall or a slice, with our world on one side and the other world on the other side. This is not correct. Both worlds occupy the same space, separated by the Boundary. This is as a blanket which covers the entire world: Every blade of grass, every molecule of air. This veil does drape over 3 dimensional objects, so this can be a problem when explaining how this blankets all water and all air. This veil is at least two dimensions and at most three. I was in a home which overlapped my home. For this reason, I estimate this Boundary as being around about the thickness of a hair. When one studies my research, the issue of time being faster or slower in the other realm, will be brought to mind. This cannot be, due to the veil lifting up occasionally and our realm seeing their realm. If it was a time issue, we would see nothing.

A space of empty air is three dimensions, plus one for present time. However, it is really at least ten and at most eleven. Our four dimensions, overlapped by another four and separated by the Boundary, which contains two or three. Furthermore, my theory to open time, proven for the past at least, would add another two dimensions. This would be one for past time and one for future time: The total then would be twelve or thirteen, depending on what the Boundary consists of.

The natural occurrences of our atmosphere occasionally lift the veil and we see the other world. For example, the physical flying crafts of antiquity and the physical resurrected, which are called ghosts. Would it be possible to duplicate the conditions of the atmosphere, when the veil was lifted?

A home with proven ghost sightings means another home overlaps that one. The ghosts as said before, are living physical people and they live there.

Setting up a round frame, with a diameter of six feet and one to three feet deep: The machines and equipment are attached to the frame. The air within the frame is manipulated, to attempt to duplicate the natural occurrence of the atmosphere. A great effort is made, to what has never been tried among man, a _"DIMENSION SEPARATION VIEWER."_

Wheels must be on this unit, for this may not be a window, but only a slice. Moving this around the room, one will attempt to see furniture and decor of the home that overlaps theirs. If this occurs, the formula will be a success. The main goal however, is to see or talk with the people across the "Realm Boundary." The result being, communicating with the dead of our world, who are resurrected in that world.

An invention worthy of treasure and spirit!

Johnny Vincento

UPDATE: While looking through the King James Bible I ran across this amazing verse.

"And God said, Let there be a firmament in the midst of the waters, and let it divide the waters from the waters. And God made the firmament, and divided the waters which were under the firmament from the waters which were above the firmament: and it was so. And God called the firmament Heaven. And the evening and the morning were the second day."
(Genesis 1:6-8 KJV) <u>*Is the realm boundary the higgs field?*</u>

I call the separation "Veil" of worlds the "Realm Boundary." Here it is called "The Firmament." In the original Hebrew of the Old Testament that word means a very thin layer.

UPDATE: The key to building the "DSV" machine is "Vibrational waves." Each world has its own vibration and light. The light doesn't bother a person who is Resurrected or in other words Teleported. Because the Rematerialization of the flesh is intune to that worlds 'Vibrational Matter" and light. However, in our world we are not in tune to Heavenly light. Paul the Apostle in the Bible went blind looking at the light from the other world. Anyone getting close to making an operational "DSV" must wear heavy dark sun

glasses.

What else can be used? Magnetic waves or influences. Eleco waves or influences. Combination of both, or even a very high turbo speed fan to manipulate the air. Possibly water as well, putting glass on each side of the frame manipulating the water inside to reveal the hidden images. The following may give further information on the different laws of physics while making the "DSV."

In the fifth month of the year 2014 on the 16th day. This being a Friday. Sitting alone in my office chair with the lights off and the the sky becoming dim. I was in distress   because of my present situation and happy too thinking of what the voice meant. About two weeks before while putting on my "Mary with baby Jesus" necklace: A verbal voice in my right ear said "This is going to be the best year ever."

I am alone with my thoughts on that May 16th with my back to the wall in my desk chair. There is about a foot distance separating my chair from the wall. I hear above me! Electricity! "cheee cheee --- cheeeeeeee cheeeeeeee --- CHEE!" With this sound came a bright white light that illuminated the room. Then with the last "CHE!" sound the entire room was so bright that this type of light doesn't exist on Earth. It was so bright that I was thinking "Don't look to where it is coming from." I thought I would get blind if looking directly at the light.

This opening came from about a foot and a half above my head: While I was sitting. The wall was behind me about a foot. So somewhere in that area. Out of the air or wall came hovering straight across the room an object that I will describe to you: It was bigger than a pinball machine ball. And smaller than a golf ball. In-between those two was the size. It looked like it was filled with liquid Mercury. At the same time had "Ruby blue" random spots like our Earth would look like if all the land was Ruby blue in color and the rest of the planet was liquid Mercury. This didn't look anything like a mini Earth. I am describing the randomness of the color Ruby blue.

The device cruised very slow in a straight line about 18 feet and then disappeared.

I seen spirits and they look to be filled with liquid water. They are also larger. This it seems was a drone of some sort sent through dimensions to visit or view me. The substance may have had a seeing device inside. Why else would it bother to cross into our dimensions? That is a lot of effort to do such a thing. If it wasn't a lot of effort, things like that would happen all the time.

It was enclosed in something clear, unless a liquid ball full of Mercury and what else the Ruby blue was could be made by using a magnetic field.

If it wasn't Mercury then it was silver metallic liquid paint. I doubt it was that. So that's why I am so sure it was Mercury or something that looked the same.

After it disappeared I felt the wall and it was not warm or hot or damaged in any way. No sign of it coming from the wall. So it had to have opened dimensions through the air itself.

The following was written and shows a bit of understanding.

Jesus said, "Images are visible to people, but the light within them is hidden in the image of the Fathers light. He will be disclosed, but his image is hidden by his light." Gospel of Thomas: 83

Replacing the word "Fathers light," "He" and "His" with "OTHER WORLD" tells us something about different dimensions and seeing those images. Plus adding "The Images" to them makes this very heavy duty Jesus saying understandable.

Jesus said, "Images are visible to people, but the light within (the OTHER WORLDS images) is hidden in the images of the (OTHER WORLD). (THE OTHER WORLD) will be disclosed, but

Johnny Vincento

(the OTHER WORLD) images are hidden by (the OTHER WORLDS) light." Gospel of Thomas: 83

Images are visible to people,

but the **light within the** other worlds **images** is hidden

in the images of the other world.

The other world will be disclosed.

but **the other worlds images**

are hidden by the other worlds light.

THIS IS MUCH EASIER TO UNDERSTAND EVEN FOR ME.

Images are visible to people,

but the other worlds images are hidden by the other worlds light.

That seems to be a description of what I seen. The image of the device was seen as long as the high intensity light from the other world shown on it. As soon as the hole or opening closed the image disappeared. The other light no longer was

there to make it visible.

The DSV is not about duplicating the light of the other world to see the images. However, that could also be used to influence the empty air within the frame. No matter what else is used, the key to creating the DSV is using "Vibrational Waves" along with any other influences needed.

The main objective of the DSV machine is to duplicate the natural occurrences of our atmosphere that lift the "Veil" --- "Firmament" --- "Realm boundary" to reveal the hidden images of the other world.

LETTER TO THE UNBORN OF THE 22ND CENTURY: THE THEORY OF BENDING SPACE BY THE REMATERIALIZATION OF MATTER THROUGH DIMENSIONS

The laws of dimension show that mortal existence can never simultaneously dwell with immortal existence. They however, can visit each others realms for short amounts of time. Space can be bent on Earth. This is due to the different laws of physics of the other realm. One can be at point A in the mortal world, cross through dimensions, and rematerialize at point B in the second world. This being accomplished, because of the "Theory of Matter." If one combines the scientific communities "String theory," with my "Theory of Matter," a complete picture is shown of new laws of physics.

Applying these principles past the gravitational pull of the Earth, an experiment can be made. <u>*This will be an attempt to travel the universe by bending space, instead of by the conventional means.*</u> A space craft with a spiritnaut in his suit, must leave the pull of the Earth into empty space. The "Law of duplication" shows that ones clothes are duplicated along with the flesh. After fasting for the proper amount of days, the fatigue is implemented. The mystic astronaut closes his eyes: Where will he rematerialize to? There is a good four possibilities:

1. If the "Theory of Matter" applies to empty space, then this means there is two realms of overlapping vacuumed space. One for the mortal, the other for the immortal. The planets and stars should not be different in either realm, only the

viewpoint of realms. This however, is unknown territory in the study of dimension. Since the vacuum of space is involved, the spirit may duplicate the entire environment along with the spiritnauts clothes and flesh. So it would be a ship rematerializing in the exact same space overlapping the other ship. Just as in "Journey to my father," being in the home that overlapped mine. The crew would not be duplicated.

2. The spiritnaut may rematerialize on the surface of the second Earth.

3. If The "Theory of Matter" applies to a second realm of empty space, the entire ship and spiritnaut may rematerialize to another part of the unknown.

4. If the "Theory of Matter" does not apply to empty space, regardless if there are two realms of vacuumed space, then the possibility of rematerializing on another planet with an immortal realm, must be considered. Space can only be bent by visiting an opposite realm. One can never bend space and visit the same realm.

Since Earth is a dual planet, others must exist too. It doesn't make a difference how far they are. That is what makes this theory unique. When one bends space, distance is not a factor.

Now a planet can be dead and possibly still have

life. If every living thing were destroyed on Earth, there would still be life by crossing dimensions. Applying the same physics equations to a dead planet as was applied to Earth may uncover a dual planet. It would be interesting to undergo the "Teleportation" on the surface of Mars. Through the mystic art of time travel what would one see? The duplicated image of the world, the way it was billions of years ago? If there was any type of life forms and water and plants, they may still exist. These are the questions that will arise when one starts to explore the universe.

The mystic astronaut transcends the universe, space and time.

**LETTER SENT TO NASA
IN THE YEAR 2006**

**NASA
Department for the man
Mars mission:
300 E Street S.W.
Washington, DC**

Operations:
The enclosed manuscript is eight years of research. It appears to be religious and that is how this work started. However, during the research, <u>one new law of physics after another, along with the formula to cross dimensions</u> shown themselves.

Please review this compilation of work and do not look at it as a religious text. Look at through a physics point of view. <u>Religion is physics!</u> The word "Resurrection" is a religious word, however, the word "Resurrection" really means, "The rematerialization of matter through dimensions."

So when one reads of religious events of the New Testament for example, these are not religious, they are an individual harnessing the different laws of physics that are natural in the world that overlaps ours..

Combined is religious terminology with physics terminology, but it is really all natural laws of interdimensional physics.

PROPOSAL FOR NASA TO IMPLEMENT A STUDY GROUP OF 10 VOLUNTEERS, TO ATTEMPT CROSSING DIMENSIONS

This study will only take one week, attempting to activate the formula on page three. After the study test is complete and they say the same thing as the research enclosed, then the <u>"THEORY OF BENDING SPACE BY THE REMATERIALIZATION OF MATTER THROUGH DIMENSIONS"</u> must be attempted.

The benefits greatly outweigh a one week study test.

Humanity will never travel the universe unless he understands the spirit. Once he knows how to harness the spirit, then he will know dimensions. And once he knows dimensions, then the universe may be able to be traveled.

LETTER TO THE READER

The first will be last and the last will be first. A person who lives a righteous life and prays everyday will be last to enter the Kingdom of Heaven. For Heaven is the highest purest world out of all of them. Even looking at the sky give you a high of happiness.

The ones who are paralyzed, blind, those who lost a pet, wife or a child, the ones who have spirits that are crying in pain, because of their mortality: Those are the ones who will enter the second world first. They have nothing to lose and everything to gain. Even though they are afraid of taking the Journey, they will, because they have to know.

The one who was born blind should be able to see, the one paralyzed will walk, and the one who was afraid of death will find life.

Their spirits will rejoice, they have seen the end and it is a beginning. Along with the new world comes a functional body. For that few minutes through the "Silver Cord," they will see and experience what awaits them.

Only those who attain the "Teleportation" while they live, will be guaranteed the resurrection at death. All others will have to take their chances.

Anytime a unique individual discloses something, that's a hundred years ahead of its time, the world considers him mad.

But isn't that the way it's always been?

Johnny Vincento

LETTER TO CATHOLIC PARENTS

The Lutheran private school is superior in many ways over the Catholic institution. Martin Luther reformed the church in the 1500's, due to the many problems he seen. One imparticular, was the selling of indulgences. This was a con on the people to get their money. The indulgence was a piece of paper that said, if you buy this, your dead relative will be taken out of Hell or Purgatory and brought to Heaven. A Lutheran church is a Catholic church reformed. The following are the changes and the reasons to take your children out of a Catholic private school and into a Lutheran private school.

1. You do not have to confess your sins to a priest, for there is no mediator between your Father in Heaven and you.
2. You are not judged on your works, but on your justification of faith. An evil person can do a good thing, but that doesn't make him good. A good person can do an evil thing, but that doesn't make him evil. Instead, one is justified by what is in their heart. One must do good not for the work, but for what is in their heart.
3. Your children will be in a safe

environment, away from those priests who look at your children with vile thoughts. And away from those priests who act upon their thoughts.
4. Have you ever heard, in your entire life, of a representative of the Lutheran church or school being arrested for anything?
5. Women are allowed to preach.
6. You do not have to call a man Father, when you already have a Father at home and a Father in Heaven.
7. Most of the churches and schools still have the same Gothic style buildings: Stained glass and all.
8. Tuition is more cost effective at a reduced rate.
9. Students wear uniforms similar to the Catholic schools
10. Lutheran representatives are allowed to have spouses and children.
11. The Bible is studied the same, with the exception of this reform that I just stated.

12. The Catholic church sprinkles water to Baptize which is incorrect biblically. Nowhere in the Bible does it say that. The proper way is a full

submersion when the child is old enough to make his own choice: Not when they are babies.

Families do not know this information, that's why they stay with what they were brought up with. So transfer your children to a clean uncorrupt Lutheran private school environment.

The world does need the Catholic church for "Rights of Exorcisms." There is no corruption in this, for only a priest who has a pure heart will be successful. If a corrupt vile priest attempted such a thing it would fail: Because that priest is as bad as the demon spirits he's attempting to remove.

LETTER TO THE PEOPLE OF ASIA: BUDDHIST, HINDUS, CHRISTIANS

Greetings:
I give you a new path to obtain Enlightenment. You are the most adored in my heart of all people. I am not Asian, but I am you. My research now perfects the skill to experience pure Truth personally: Seeing with your own eyes.

If you wish to see "The Gospel of Thomas #84?" Jesus said, "When you see your likeness, you are happy. But when you see your images that came into being before you and that neither die nor become visible, how much you will have to bear!". Ask your Creator on your Journey to show you.

To get rid of karma payback, I describe here how I did it:
I was Baptized on December 4th, 2016 and I was reborn so the old me died. Before I went under the water I said

"OK this is it, I repent for all my sins, I take Jesus Christ as my Lord and Savior. The karma is over, I cant take it anymore, there is no more payback. I also repent for any sins I may have done in any previous life if I even had one."

Then I went under the water of full submersion. And this is how I removed my karma payback. Don't ever get Baptized with sprinkles. Nowhere in the Bible does it say sprinkles is how to Baptize. Everything in the Scriptures say "Full Submersion."

You may think that going on a Journey to Heaven will get rid of karma: It doesn't. However, you can ask your Father in Heaven to erase all your karma payback. There must be repentance. As for me, when I say born again, I refer to what I did. I died in my sins. The old

Johnny Vincento

me is dead and my sins were washed clean, for this life and any previous lives (if I had any). When I came out of the water I was reborn and the karma was gone.

Before I washed my karma away it was at its worst. I found myself being falsely arrested. Sitting in the back of a Police truck with another prisoner. Our hands were handcuffed from behind and a cage was in front of our faces. The prisoner next to me said "Do you believe in karma?" I said "Yes." He replied, "Maybe this is payback for what we did in the past." I answered "The question is: How much karma do we have coming to us?" Before I finished the word "Us", a car purposely steered directly head on toward the Police truck. We were on two wheels. There was no one on the road at 3AM. Just us and the girl driving the car.

She didn't know what happened: BUT WE DID!

X

FINAL THOUGHTS FROM THE AUTHOR

This work is my gift to the world. Many people write worldly things for profit or fame. Putting their names so big on the cover, that one cannot even find the title of the book. Seeing a page describing one of my Journeys, it is just a page to you. However, what you do not know is, I risked my life to gain what is on that page. I never knew if I was going to come back. I knew there was a risk and I took it to explore new worlds.

Through consistent Journeys, I can tell you that the Journey is safe to take. We were meant to take this Journey to ensure our resurrection. At the time I knew not this. Now you know through my risk, that the method is safe to accomplish. Many people write on opinion, which is worthless, and explain what they think a Gospel passage means. This is worthless for Enlightenment. I have enclosed the method to my claims, for anyone who denies my truth is welcome to call me on it and see with their own eyes.

Johnny Vincento

Anyone can write and talk about worldly things through opinion. Only one unique can write and talk about things, without opinion, that are not of this world!

I do not consider myself as an author. I am a "Messenger." However, in the age of the unborn, people will consider me many things. There are no better words to describe me than a "Pioneer Applied Physicist Spiritnaut." I am self taught and even if I went to the greatest Science University in the world, they wouldn't help this research and development one bit. I have to make up new words because Theoretical physics doesn't describe my "Teleportaiton Research." Furthermore, this Teleportation is not a theory and works every time. Those who are bold enough can seek to obtain knowledge from the other realm and profit from that knowledge.

Then there is the person only known as "The Greedy One." With my work in hand, he will study it and practice making physical objects appear in another world. Then when he perfects his Art in that existence, he will attempt to make gold bars on Earth. For this ones desire is to be the doorway between worlds, harnessing the physics of that world as did Jesus Christ. Whether he is successful I was not informed. This information was implanted in my mind after the last Journey. So after some time, I wrote it down, because it lingered in my thoughts as if someone down loaded it into my mind. When this one arrives "The New Age of Enlightenment" is at hand.

Now no less than five times I should have been killed. Each time was a divine intervention. Some examples would be: When I was with my grand parents and father at the family business in the ghetto. Two young black gang bangers pulled up and yelled "Let's do it!" They opened the two rear doors to get rifles from the back seat. I never thought I would make it to the office and was expecting a shot in my back. I should not have been there that day. However, I foiled their robbery when I got to the office gun and they aborted their mission. Another example was when I had a thought to put on my bullet proof vest. That thought was my own so I ignored it. Then a thought not my own said, "You should go back upstairs and put your vest on." That night I hit a tree. The police said that I had bent the steering wheel and I had hit the tree going around 60 mph. The vest I had on was equipped with a shock plate. Then there was another time when I was being robbed with a knife to my throat. An actual voice yelled within or without my head, "RUN!" That was the only time I ever heard what the Holy Ghost souded like.

UPDATE: This is 20 years of research and since then I have had many times the Holy Ghost tell me things in a verbal voice. All were helpful and guiding for my prosperity. One example was when I put on my baby Jesus and Mary necklace. A voice in my right ear said "This is going to be the best year ever."

As far as Angels go. I can tell you that they really exist. When I was very sick I woke up a little after the 11pm hour and seen two Angel girls kneeling, facing each other, hands together in prayer, with their heads bowed down. One was a blonde white girl with short straight hair and the other was a darker skinned girl with short black straight hair. They were on my left and I turned to my right and on the end table was another female Angel sitting on it half way. Then I said out loud "At least someone loves me." and then I went back to sleep.

An individual must visit the other world if one truly wants to be happy in life. That one will not have fear of death, for what awaits will be shown. For this reason, there are no greater people, than the American Indians of the 1800's. Of course I am not talking about the bad individual indians of History: Overall. A right of passage was taken by young Indians and called, <u>"A Spirit Journey."</u> It involved going into the woods, alone, for four days and fasting. On the fourth day, dancing in the rays of the sun until passing out.

To make this work complete I would have four questions to ask.

> *1. How can the future be seen if it didn't happen yet?*
>
> *Update:*

Our time line plain goes into space and bends like light. It goes over future and past timelines. When one sees the past they are seeing the present. And when sees the future they are seeing not the future but the present.

2. What planet holds similar air and life as ours in the galaxy?
3. How can a human visit that planet?
4. How can one make the "Dimension Separation Viewer" so there can be a window between Heaven and Earth?

Even if one achieves the enlightenment, he still will run into animals in human form on this Earth. For this reason, it is always best to use your education of your past and use the law for your favor:

Johnny Vincento

NOT EVERYONE MAKES IT (SONG)

<u>*Come take me by my hand son*</u>
<u>*Come take a walk with me*</u>
<u>*I'm glad you have made it son*</u>
<u>*The trips even hard for me*</u>

It took me days of fasting
Three days to be here with you
<u>****Son not everyone makes it!****</u>
<u>*Unless they follow you through*</u>

The road not taken is the road I'm taking for you
My eyes have eyes which see what others can't see
I walk in body which waits for my body for me
The road not taken is the road I'm taking for you

Have you seen my companion!
Will my spirit you mend?
Do you ever think of me?
I need to see her again

<u>*You need not to worry son*</u>
<u>*I have her waiting for you*</u>
<u>****But not everyone makes it!****</u>
<u>*Unless they follow you through*</u>

The road not taken is the road I'm taking for you
My eyes have eyes which see what others can't see
I walk in body which waits for my body for me

The road not taken is the road I'm taking for you

> *Do you ever get lonely?*
> *I will always be with you*
> *Will you ever leave me!!!*
> *I am always here with you*
>
> *You need not to worry son*
> *I have her waiting for you*
> ****But not everyone makes it!****
> *Unless they follow you through*

The road not taken is the road I'm taking for you
My eyes have eyes which see what others can't see
I walk in body which waits for my body for me
Coming into being guarantees me being with you
Healing my body and healing my spirit for me
The road not taken is the road I'm taking for you

> *The song is about seeing my dog*
> *once again who died.*

Johnny Vincento

Letter to all suicide killers and evil ones

BEFORE YOU DO SOMETHING EVIL OR STUPID TAKE THE JOURNEY. YOU ARE MOST LIKELY GOING TO DO SOMETHING EVIL BECAUSE YOU THINK THAT THIS WORLD IS ALL THERE IS. IF YOU TAKE THE JOURNEY AND GO TO HELL YOU WILL COME BACK AND BE A SAINT. BECAUSE YOU WILL WANT TO CHANGE WHERE YOU ARE HEADED. IF YOU TAKE THE JOURNEY AND GO TO HEAVEN LEVEL ONE OR TWO, YOU WILL COME BACK AND BE A SAINT AS WELL. BECAUSE, YOU DON'T WANT TO RUIN WHERE YOU ARE GOING. AND IF YOU GO TO THE THIRD HEAVEN YOU WON'T LIKE THAT SO YOU WILL ALSO COME BACK AND BE A SAINT.

AFTER YOU TAKE THE JOURNEY IT WILL BE GOOD FOR YOU TO BE BORN AGAIN AND BE BAPTIZED. I REPENTED FOR ALL MY SINS AND I SAID ALOUD IN FRONT OF THE CHURCH PEOPLE. "I TAKE JESUS CHRIST AS MY LORD AND SAVIOR. I REPENT FOR ALL MY SINS. EVEN SINS I MAY HAVE DONE IN A PREVIOUS LIFE. IF I EVEN HAD ONE. THE KARMA IS GONE. THERE IS NO MORE PAYBACK. WHEN I GO UNDER THE WATER I AM REBORN!"

JOURNEY TO THE THIRD WORLD THROUGH "THE SILVER CORD WORMHOLE"> THE THIRD HEAVEN (LEVEL 3)

AND IT CAME TO BE THAT I WAS TAKEN AWAY TO THE THIRD LEVEL OF HEAVEN. I WAS THERE IN THE SPIRIT AND IN THE BODY. I WAS LOOKING OUT OF MY OWN EYES AND THE REALITY WAS EXACTLY EQUAL TO THIS WORLD. TOTALLY CONSCIOUS. NO HAZYNESS BUT CRYSTAL CLEAR.

THE VIEW WAS STRANGE TO ME. FOR I SEEN PEOPLE OF HALF ANIMAL. ONE WAS THE SIZE OF A MAN BUT WAS A CAT. IT LOOKED LIKE A BLACK AND WHITE FUR SUIT WITH A TAIL AND YET WAS A MAN AT THE SAME TIME. THE OTHER WAS ALL BLACK FUR EXCEPT HIS FACE AND WAS A MAN WITH A TAIL ALSO. THOSE TWO WERE PLAYING SOME TYPE OF GAME ON A ROUND TALL TABLE LIKE IN A BAR. THEY WERE TO MY RIGHT ACROSS FROM THE BANISTER NEXT TO ME. THE TWO WERE ON THE OTHER SIDE OF THE ROOM ABOUT 30 OR 35 FEET. ONE WAS STANDING ON THE LEFT (BLACK FUR) AND THE OTHER SITTING ON A TALL STOOL (BLACK AND WHITE FUR). LIKE IN A BAR

WITH THE TALL STOOLS AND TALL TABLES. THEY HAD EARS NOT LIKE A CATS BUT BIG EARS ON THEIR HOODS OR HEADS LIKE A LIONS. I ASKED ANOTHER MAN IN HIS THIRTIES AND HE WAS HUMAN. "IN THIS WORLD ARE YOU ALLOWED TO HAVE SEXUAL RELATIONS WITH A WOMAN?" HE SAID "NO THAT RARELY HAPPENS, THE ONLY WAY THAT COULD HAPPEN IS IF THEY CAME HERE AND WERE ALREADY MARRIED." SO I CONTINUED TO WALK AROUND IN THIS LARGE HOME AND CAME ACROSS THREE OTHER MEN. AND I SEEN NO WOMEN THERE FROM ALL THE PEOPLE I SEEN IN THE HOME. I SAID "WHERE AM I?" THE ONE GUY IN HIS TWENTIES SAID "WHERE DO YOU THINK YOU ARE?" AND I REPLIED, "I THINK I DIED." THEN THE THREE ATTACKED ME AND PUSHED AND HELD ME AGAINST THE WALL. THE ONE IN HIS TWENTIES PUT A DOUBLE BARREL 12 GAUGE SAWED OFF SHOT GUN IN MY FACE AND YELLED. "OH YOU WANT TO DIE MOTHER F---ER!! NOW I KNEW I COULDN'T BE KILLED BECAUSE I WAS IN THE SPIRIT, AND THE BODY: AND I WAS IMMORTAL. BUT I DECIDED NOT TO SAY A WORD. I DIDN'T WANT TO SEE WHAT WOULD HAPPEN GETTING SHOT IN THE FACE. I THEN CAME BACK FROM MY JOURNEY.

PAUL WRITES WHAT HE SEEN AS WELL.
2 CORINTHIANS 12:2-4 KJV

"I KNEW A MAN IN CHRIST ABOVE 14 YEARS

AGO (WHETHER IN THE BODY, I CANNOT TELL; OR WHETHER OUT OF THE BODY, I CANNOT TELL; GOD KNOWETH;) SUCH AN ONE CAUGHT UP TO THE THIRD HEAVEN. AND I KNEW SUCH A MAN, (WHETHER IN THE BODY OR OUT OF THE BODY, I CANNOT TELL; GOD KNOWETH;) HOW THAT HE WAS CAUGHT UP INTO PARADISE, AND HEARD UNSPEAKABLE WORDS, WHICH IT IS NOT LAWFUL FOR A MAN TO UTTER."

SOME PEOPLE TALK ABOUT GOING TO HELL ON TV PROGRAMS. SOME DESCRIBE HOWEVER, THE THIRD LEVEL. BECAUSE THAT IS WHERE ALL THE MEAN ANGRY PEOPLE GO THAT ARE NOT BAD ENOUGH TO GO TO HELL.

(Paul is talking about himself most likely.) Did you ever wonder why the Sphinx is in the image of a cat man? Thoth made it. He may have seen this world with the cat people. To really get your mind going, another face was found on Mars: A woman with cat ears. Like a lioness has. One looks broke, but the fur is there too.

XI

VISION FROM THE SPIRIT OF TRUTH "2030 AD"

In the year 2013 about a month after the Russian asteroid exploded over head and another asteroid flew by earth the same day: A vision was given to me. After two weeks of continual prayer to the Spirit of Truth, as written in the Gospel of John. I was taken away to the future which was the present. I found myself looking out of my own eyes in a wide tunnel vision type sight. Blurry around the ends and clear in the center. I was sitting on a warehouse floor with my back to a wall. To my right was a doorway that was open. It was so tall that a train could have easily fit in. However, I seen no tracks. Across the warehouse I could see a large yellow forklift, a blond woman and about two other men. But the edges of my sight were blurry like I said. I also seen at least one skid. The fork lift was the type you sit on and drive. It was larger than the propane standard dock jeeps. Probably gas or diesel. It had a black seat. There was a man sitting to my right with his back against the wall as well. He said "This may be the extinction of the human race, we only have enough food for three days, we need to get to the Southern states!" We both stood up and I replied "What year is it?" The man said "2030." I said "That's right, last year was 2029." Because, I was there in the present just as if you were saying what last year was. And again I say I was there. I was an active participant in being shown things to come.

Then the vision gave me a tour of the rural town on the back of a motorcycle. The guy was driving and I was on the back looking at an abandoned town. The debris was everywhere and we passed a light pole that was knocked at about a 60 degree. Not 45 degree that would be too far. No buildings were taller than 3 stories. The people just got up and left. The sky was nice and the weather too. So whatever that guy was talking about didn't happen yet because the sky was nice and clear.

Then we went into a shop that appeared to be a pet store.

Johnny Vincento

The glass door was open and the inside looked ransacked. It had stairs going to the upper floor. In front of me was a 6in high clear plexiglass box about 3ft x 3 ft. but there was a taller box that fit inside to the halfway point. Making a waterfall over its curved slope. Inside the water was a sliced dead fish and I could smell the fish. I put my hands in the water and tasted it. I could feel the water and it tasted good but as of a fish. I called out! "Hey I found a water source, now all we need to do is find some food and we will be all set!" after that I went upstairs. It was ransacked too. On a dresser that was brown I seen a strange pair of pink glasses. One side was normal and the other lens was as a window with blinds. Probably sun glasses.

Then I heard a sound from the front of the building and I never heard such a sound. It sounded like a machine or a drone. It went "UMMMM-MMMMUM----UMMMMMMMMUM----UMMMMM-MMMUM----UMMMMMMMMUM. Then I was taken back to this world and time.

Be prepared with food water and supplies for you and your family. The weather and sky were nice so at least by the Winter months of 2029 supplies should be purchased. If something happens it will be after the first quarter of 2030. I PUT MY WHOLE REPUTATION ON THIS VISION.

One other note: I tried to find the name of the asteroid by using automatic writing. That is very unreliable, but I did come up with the name: "CA-LANDRA." The only reason I am adding this is because I never heard of such a name. And when I looked a few months later at a list of 10,000 asteroids, that name was there.

I in no way put my reputation on any automatic writing. I only add this just in case and because that is the biggest coincidences I ever seen.

It is written of a watchman in Ezekiel 33:1-6 KJV "Again the word of the lord came unto me, saying son of man, speak to the children of thy people, and say unto them, when I bring the sword upon a land; if the people of the land take a man of their coasts, and set him for their watchman:
If when he seeth the sword come upon the land, he blow the trumpet, and warn the people; then whosoever heareth the sound of the trumpet and taketh not warning; if the sword come. And take him away, his blood shall be upon him. But he that taketh warning shall deliver his soul.
But if the watchman see the sword come, and blow not the trumpet, and the people be not warned; if the sword come, and take away in his iniquity; but his blood will I require at the watchmans hand."

Johnny Vincento

THE THREE CROSSES IN A ROW UPRIGHT WILL BE OUR SYMBOL.

WEAR IT ON YOUR CLOTHES OR SKIN AND WHEN YOU SEE ANOTHER WITH THE SYMBOL YOU WILL SMILE.

MANY FRIENDSHIPS, MARRIAGES AND GROUPS WILL BE FORMED IN THIS WAY.

When I was in "Another Land" I seen a girl who didn't have any hair. I gave her hair by wishing and focusing on it with my right hand out. And when the hair materialized on her head: It had three small silver crosses hanging from it.

TO ALL EXPLORERS!
ALL ADVENTURERS!

I CALL ON YOU!!!

AS
SCOUTS TO OTHER WORLDS.
THE JOURNEY IS SAFE TO TRAVEL.
VISIT YOUR STAR BROTHERS & SISTERS!
HAVE QUESTIONS TO ASK THEM.

COME TO MY WORDS
ALL THOSE WHO'S SPIRITS
ARE CRYING OUT
IN PAIN.

I AM THE MESSENGER

To contact:
johnny.vincento@yahoo.com